Praise for *The Chemical Detective*

'Action, intrigue and a stonkingly modern heroine. It's a blast.'
Sunday Times Crime Club

'Imagine the love child of Jack Reacher and Nancy Drew…a delicious cocktail of dating and detonations. Call it Mills and Boom.'
Evening Standard

'Intricate, seductive and thrilling. Erskine's writing glows with wit and danger. And can't help reminding you, with every page turned, of how close we all are to detonation.'
Ross Armstrong, author of *The Watcher*

'Fiona Erskine is an engineer, and in Jaq Silver, who shares her profession, she has created a wonderful antidote to all the resentful, floppy victims of much domestic noir.' *Literary Review*

'A stunning, cinematic debut that's going to land on the 2019 thriller scene like a half-kilo of silver fulminate.*
*stuff that goes bang'
Andrew Reid, author of *The Hunter*

'Bring it on! …dangerously thrilling.' *The Booktrail*

'An audacious, female-led thriller which took the disposable women of the James Bond franchise and flipped the concept entirely on its head.'
Chemistry World

'Erskine weaves her tale of suspense into a first-class thriller, the drama escalates with terrifying authenticity. A must-read for engineers.'
Nick Smith, *Engineering & Technology*

'A complex and increasingly addictive story that soon becomes a page turner.'
The Chemical Engineer

FIONA ERSKINE is a professional engineer based in Teesside, although she travels often to Brazil, Russia, India and China. As a female engineer, she is often the lone representative of her gender in board meetings, cargo ships, night-time factories and offshore oil rigs, and her fiction offers a fascinating insight into this traditionally male world.

THE CHEMICAL DETECTIVE

Fiona Erskine

A Point Blank Book

First published in Great Britain, the Republic of Ireland and Australia by Point Blank,
an imprint of Oneworld Publications, 2019

This mass market paperback edition published 2020

A CIP record for this title is available from the British Library

ISBN 978-1-78607-714-1
ISBN 978-1-78607-493-5 (ebook)

Typeset by Geethik Technologies
Printed and bound in Great Britain by Clays Ltd, Elcograf S.p.A.

Oneworld Publications
10 Bloomsbury Street
London WC1B 3SR
England

For Marion and Norman

PRELUDE

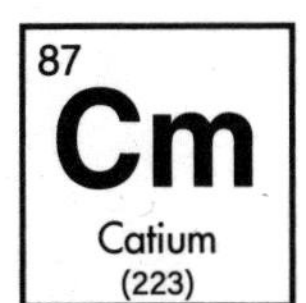

Thursday 24 February, Teesside, England

The trouble with Semtex is the smell. Dogs can sense it. Most humans can't. Boris could. Not the plastic explosive itself, you understand; neither RDX nor PETN – the main components – have much of an odour. The scent comes from the tracers added, to make sure it doesn't fall into the wrong hands. Hands like his. Chemist's hands. Wide hands with long fingers, calloused from handling hot glassware, thickets of black hair curling over the knuckles, little copses between the joints. Hands now gripping the steering wheel of a five-axled articulated lorry hurtling towards the Zagrovyl factory in Teesside.

Boris only carried a small amount of Semtex these days, just enough for his own personal use. He kept it in a Tupperware container, wrapped in cling film, under his sandwiches. Sentimental value, really. He'd moved on. To some it might look like a backward step, from laboratory shift work to long-distance lorry driving. But only to those who didn't know the tedium of analytical testing. The same samples, the same tests, the same results, hour after hour after hour. Not like the old days, when you had thorny problems to solve and real fires to fight. Nothing more boring than a well-run factory. He was glad when they sacked him. Glad to be free of the monotony. Glad to be out on the road. These days, his insight into tracers was a key skill for the job.

Boris yanked the wheel to the left and hauled the artic into a lay-by with a view. The chemical plant skulked on the far side of a silver fish-scale fence. One factory was much like another. Plumes of steam billowed into the sky, glowing orange in the sodium lights, bright against a dark winter day. He traced the familiar shapes in the condensation of his side window: an hourglass – the cooling tower curving to a waist and then flaring out again; two thin vertical lines – the nitric acid absorption columns lit up like Christmas trees; three circles – the ammonia storage spheres, massive metal balls trapped by sturdy legs to stop them rolling away; a rectangle – the ammonium nitrate prilling tower looming over the A19, the main road out of Teesside.

The wind whistled up the river, screaming through the gap between the warehouses, bringing with it a faint whiff of sulphur, reminding him of home: Pardubice in the Czech Republic. The Semtex factory where he trained.

He watched the car park from the lay-by, waiting until the last company car roared away, before driving up to the gatehouse and presenting his papers. At the collection bay he plugged a small black box into the vehicle's lighter socket. It bleeped and flashed, a red light showing it had located the Zagrovyl computer network. He tucked the jamming device under the passenger seat before turning off the ignition and stepping down from the cab.

'Snow Science, right? Two tonnes?' The bald warehouseman tapped his keyboard. 'Bloody system down again.'

Boris slid his papers though a hatch. 'Twenty tonnes.'

'Fertiliser grade?'

'Technical grade.' Boris jabbed his finger at the product code on the order.

'You sure?' Baldy frowned and inspected the order line by line. He picked up a phone, running a hand over his eggshell-smooth head as he waited. When there was no response, he shook his head and cursed. 'Lazy tossers, all buggered off early.' He slammed the receiver back into its cradle. 'I'll get you loaded up in a jiffy, mate.'

The metal ramp screeched against the concrete floor as a forklift truck drove into the back of the lorry, delivering the first pallet. Two forklifts worked in tandem, an intricate dance, weaving and turning on a sixpence as they loaded the cargo.

Within fifteen minutes it was finished. Fast and skilful, these old men of the north.

Boris secured the load, signed the paperwork and drove out of the factory gate.

Click. Location 54.597255, -1.201133. Intensity 800X

Instead of taking the A19 south, he headed east to Haverton Hill and a decrepit warehouse lying in the shadow of a blue bridge. A damp chill rose from the misty river. Boris shivered as he opened the cab door and scanned the quayside.

A tall, thin man materialised out of the fog, moving slowly with laboured, jerky movements. He emerged into the sidelights: dark coat, spiky black hair, gaunt white face. The Spider. Christ, this run must be important.

'So?' The question came out as a hiss.

'All good.' Boris pointed to the trailer. 'No problems, boss.'

The Spider pressed a button and battered doors began to open, groaning and squealing with neglect.

Boris backed the lorry into the warehouse and hopped down from the cab. 'How long will it take?' he asked as he unlocked the back doors and dropped the ramp.

'Assist,' The Spider ordered. 'Time is of the essence.'

Two hours later, Boris's arms ached as he manoeuvred the artic onto the southbound motorway. Bloody amateurs. Leaving him to do all the heavy work.

Boris made good time to the south coast, skirting London after the rush hour. Transport of explosives was not permitted in the Channel Tunnel, so Boris and his lorry boarded the ferry to France.

Click. Location 51.12646, 1.327162. Intensity 152X, 648C

He stood on deck, sipping at a watery English coffee, as the white cliffs of Dover receded into the mist. Plain sailing from here. He shivered as the towers of the titanium dioxide factory beside the Port de Calais hove into view, and returned to his lorry.

Click. Location 50.96622, 1.86201. Intensity 152X, 648C

The drive through France was uneventful as far as Strasbourg, but a young border guard flagged him down at the crossing into Germany for extra checks. So much for a borderless Europe. Boris remained calm. It had happened before. Nothing to worry about.

The ginger-haired guard puzzled over the papers, wrinkling his brow. 'You do know what you've got in there?'

'Yes.' Boris lied easily now. After the first few runs he knew how unlikely it was that anyone would check. And even if they did, what would they see?

Ginger picked up a phone and moved out of earshot. After a few minutes he marched back. 'Drive carefully.' He waved him on his way.

Click. Location 48.5857412, 7.7583997. Intensity 152X, 648C

Boris drove on past Baden-Baden. After lunch, near Munich, he took a nap in the back of the cab. When he woke, the stars guided his way to Salzburg and the crossing into Austria.

Click. Location 47.7994, 13.0439. Intensity 152X, 648C

As he approached the mountains, snow started falling, wet flakes that melted on impact. A weather report on the radio warned of treacherous conditions and several inches of snow up ahead. Great for the skiers, bad for lorries full of explosives and worse. Best to cross in the morning. He slid into a lay-by. A police car drove towards him, slowing as it passed on the opposite side of the road. Boris stared into the snowstorm, craning his neck to make sure it didn't turn back.

Not that he need worry too much. The dispatch papers matched the Dangerous Goods Note. The bags had the correct hazard warnings. All the papers were faultless. None of the inspections, on any of the runs, had ever uncovered a thing. After all, who wanted to poke around inside bags of explosives? You could hide anything in there.

PART I: OVERTURE
SLOVENIA

Saturday 26 February, Kranjskabel, Slovenia

A strange bed. A naked man. And a few hours to kill before the explosives arrived. The day was looking up.

Jaq stretched, savouring the smooth cotton sheets against her skin. Snowflakes danced through a web of ice on the sloping attic window. In the dawn glow she could just make out the layout of the unfamiliar room. Two doors: one of solid oak with tongue-and-groove panelling, brass hinges and a sturdy lock; the other a flat, sliding panel leading to a modern shower room carved from a corner of the attic. A pine bed, wardrobe and chest of drawers, a leather sofa and a couple of metal stools tucked under a bench that divided the bedroom and kitchenette. From outside came the faint swishing and rumbling of a distant snowplough. Inside, the gurgle of a fridge, creaks and sighs of an old house waking up and the steady, slow breathing of the man beside her.

Jaq breathed in: musk and liquorice. And a faint whiff of nitroglycerine. Her scent on his body.

She slid backwards across tangled sheets and ran her eyes over the golden curls decorating the pillow, down the ridge of his spine to the curve of his buttocks, sturdy thighs and powerful calves. Definitely a skier. One foot hung over the edge of the bed while the other was tucked under a leg forested in fine bronze hairs. A tall, blonde skier. Athletic. And much too young for her.

She grinned as she reached for the quilt – curved appliqué ridges between her fingers, uneven stitching, not machine-made – and gently covered him. He stirred but did not wake.

The room smelt of pine resin with a hint of lemon. Clean and tidy. Well, at least it had been before last night. Her eyes followed the trail of clothes across the oak floorboards. Her coat and hat hung on a wooden peg near the entrance door, but her long boots had toppled over and lay at angles to the pashmina snaking across the floor, coiled around a scarlet bra and matching thong. There was no sign of her dress, but on the chest of drawers in the corner she could see his clothes, neatly folded on top. When had he folded his clothes? While she was asleep? Certainly not as she was undressing him.

The guy from the karaoke bar. *Nossa.* What had he done to her brains last night? She'd known he was trouble the moment she heard him sing.

Karaoke. What had she been thinking of? She loathed office parties, but her boss at Snow Science had insisted on it. Team building, Laurent said, a bit of fun. Laurent was a pillock.

She slid down the bed, covering her head at the memory of Laurent's excruciating impersonation of Charles Aznavour. *Carapau de corrida.* He'd insisted on the drinking games afterwards. Sheila and Rita had the sense to refuse but Jaq could never resist a challenge.

And then the man with the golden curls took to the floor.

The moment he opened his mouth, Jaq was hooked. His voice emerged an octave deeper than she expected. He sang with authority and passion, the pitch and cadence perfectly controlled. His voice rumbled right down the small stage, across the wooden floor, up through the soles of her feet, tugging at the tight knots that held her together, unravelling all the cords of restraint with the song. An old Russian lullaby. One she knew so well.

Had she stared too hard? Clapped too loudly? Was that why the singer with the deep voice and lopsided smile singled her out afterwards? She wouldn't have danced at all if Laurent hadn't made such an arse of himself. Sitting too close. Breathing too hard. Whispering in her ear. Escaping to the dance floor was intended to put some distance between them; Jaq always danced alone. Laurent followed her, his manbag on one shoulder, lurching and gyrating, arms outstretched in invitation to an inappropriate waltz.

The stranger interposed himself, moving between Jaq and Laurent, a subtle, sinuous barrier, increasing the separation until the drunken Frenchman found another target for his amorous attentions. Jaq danced on for a few tracks, just for the joy of the music, and then made her escape.

And there he was, outside the bar ahead of her. Waiting. Something in his eyes gave her pause, drew her in. She could have walked straight past. What was it that held her? Made her stop? The gentleness of his touch as he helped her with her coat? The deep voice bidding her *lahko noč*, goodnight? Had she imagined an inflection, an upturn, a question? There was no mistaking the smouldering fire she glimpsed before he hooded his eyes and turned away. It had been a long time since a man had looked at her with such honest desire. A very long time. And, oh, *meu Deus*, how she had missed it.

'Wait!' Her lips found his, and there was no mistaking the interest with which he returned her kiss. Gentle, searching, increasingly confident. Hot lips and strong arms. She remembered him asking but had no memory of her reply, or how they ended up at his place.

Time to face the morning after the night before. Careful not to touch him, her detailed inspection must have registered. He brushed the curls from his face and wrinkled his nose. His eyelashes fluttered, and his breath became shorter, shallower.

She slipped out of bed and wrapped the pashmina around her. Where was her bag? Dropping to her hands and knees, she spotted it under the bed frame and took it to the bathroom. The scent of lemon behind the sliding door hit her like a wave. She sat on the toilet and grasped the edge of the sink. How much had she drunk last night? When the dizziness passed, she took stock. Clean towels neatly folded on a rail, a shower, sink and toilet spotlessly clean. Had he expected company? She opened the glass cabinet above the sink. Soap, cut-throat razor, shaving mirror, shampoo, cotton buds, toothpaste, one toothbrush, dental floss. A large box of condoms, somewhat depleted after last night, but no sign of a permanent female presence. Just one tidy man.

Jaq reached for her bag. Despite her love-hate relationship with handbags, her party clothes lacked sensible pockets, and this was the least-bad option. Black with silver buckles, the fabric was lighter and thinner than leather but textured, tough and waterproof. It could be carried by the arched handle like a briefcase or, releasing three ingenious hooks, clipped onto a bike as a pannier. When carrying a laptop or other heavy tools, two wide adjustable back-straps unfurled so that she could take advantage of the padded, contoured panel for extra comfort against the spine. The pleated sides, held in shape by concealed Velcro strips, made it capacious enough for most outings. It even had two parallel zips, designed to slot over the handle of a rolling suitcase, but also perfect for carrying a snowboard.

She rummaged inside the bag for her phone, encountering ticket stubs, café receipts, coins, a set of Allen keys, a socket wrench, Maglite torch, penknife, comb, sachets of hot chocolate. Ouch! She caught her finger between the pincers of a Vernier calliper. No blood, just a scratch, but she continued her search more cautiously: hydrogel plaster, crêpe bandage, latex gloves, paracetamol, ibuprofen, neodymium magnet hook, PTFE tape, thermos flask, duct tape, ball of hairy string, condoms, fuse wire, superglue, paper clip, Blu Tack, ball of rubber bands, sandpaper, a fold-up kite, Slovenian–English dictionary, an unposted letter, multiplug, catapult, USB stick, fluorescent highlighter pens, snow goggles, earplugs, spare socks, tissues, tampons, a silver propelling pencil, tube of mints, a packet of dried apricots, a tuning fork and a green marble.

Like the Tardis, the bag was bigger on the inside.

A bunch of keys fell out, clinking against the tiled floor. Odd. She unzipped the secure inside pocket where she normally kept them and, at last! there was the phone. One missed call she had no intention of returning. Amid the dross of email, a single pearl from Emma with a long, chatty message about Johan and the kids. Not now, save for later, only one bar of battery left. No message from Snow Science. She put the phone back and zipped up the keys before dragging a comb through her hair.

As she emerged from the bathroom, the naked man sat up in bed, blue eyes fixed on her face.

'*Dobro jutro!*' He switched to English. 'Good morning.'

Now that he viewed her in the daylight, was there a shadow of surprise? If so, he hid it well. What did he see? An athletic woman, naked except for a brightly coloured pashmina and a large shoulder bag. Tall, 1.75m in bare feet with a Mediterranean complexion – brown eyes, olive skin and shoulder-length hair, dark brown, almost black, except for the hints of russet fire. Well proportioned, curvy even. His smile appeared uncomplicated, no hint of embarrassment or regret, only pleasure at finding her still there. 'I don't think we were properly introduced last night.' He held out a hand. 'Karel.'

She took his hand, smiling at the absurd formality. There was hardly an inch of each other's bodies that hadn't been stroked or kissed or explored last night, and yet the contact with his hand felt deeply intimate, sending a tingle straight to her core. Careful.

'Jaq,' she said. No second names. Polite but no promises. Civilised without commitment. 'Pleased to meet you.'

'The pleasure was all mine.' He raised the quilt in invitation.

So tempting. She hesitated and was gratified by the flicker of disappointment that rippled across his brow when she shook her head.

'Breakfast, then.' He sprang out of bed, bringing the sheet with him, wrapping it around his hips. He handed her a robe. The faint hint of musk was his. She let it envelop her and perched on a stool as he got to work in the kitchen.

'A quick cup of tea, or whatever you are making,' she said.

'Scrambled eggs and smoked salmon.'

She started to protest, but the smell of butter melting in a pan made her stomach rumble. He heard it and laughed, breaking eggs into a bowl, many more than he could possibly eat alone. When had she last eaten? She'd gone straight from work to the karaoke bar, changing from boiler suit to party dress in the lab toilets. There was no reason not to eat breakfast. No reason a one-night stand couldn't be civilised.

'Nice flat,' she said.

'Belongs to a friend. But he's working abroad.' He grinned. 'I keep an eye on things when he's away.'

He served the scrambled eggs on toasted crumpets, a thin sliver of pink salmon sandwiched above the little craters of butter, turning opaque where it touched the hot egg piled in a pyramid and topped with a sprinkle of freshly ground black pepper and a sprig of parsley from a plant by the sink. A small glass of orange juice and a bowl of tea served black, fragrant with bergamot and dark tannin. The speed and ease with which he presented two perfect covers made her curious. A singer, a skier, a chef. What else could this man do? Her eyes travelled around the room and paused at the bed. Amid the otherwise orderly space it stood out, an explosion of disarray. A surge of warmth rose through her body, and she turned her attention back to the food.

'Mmmm.' Jaq wiped her lips with a napkin. 'Very good.'

Karel bowed his head to acknowledge the compliment. 'More tea?'

Jaq shook her head. Had she overstayed her welcome? He was a young man with impeccable manners, but some awkwardness was only to be expected now. She would spare him the brush-off. He would have things to do, people to see, places to go. 'Time for me to leave.' She put a finger to her lips at his polite assurances. 'My clothes?'

'I hung your dress up,' he said and pointed to the wardrobe. 'But—'

'I should go.'

'Should you?' He moved towards her.

The glass rattled in the window above, and a flurry of hail blasted the ice clear enough to reveal a storm-dark sky. No skiing today. No message from Snow Science about the delivery. Time to kill.

Karel laid a hand on her shoulder. Warm, gentle, no hint of coercion. Only invitation. Promise. He ran a finger up the side of her neck and whispered, 'Come back to bed first.'

Her skin tingled under his warm breath. When his lips nibbled her earlobe, she had to fight the urge to grin inanely. The good food, the cosy little attic, the storm outside, the gorgeous man, the firm bed. She might regret this, but. . .

Last night she'd taken a risk, let herself go with the flow, to see where it led her. What did she have to lose? Things could hardly get any worse. Forget about the past. Forget about the future. Focus on the moment.

Focus on the pleasure.

Saturday 26 February, Jesenice, Slovenia

The lorry crossed from Austria to Slovenia at daybreak. A border guard waved it down, directing the driver to a lay-by. Boris parked and swung his legs through the door, fur-lined boots followed by jeans sliding out onto the road. Snowflakes settled on his black beard.

'What you got in there, mate?' The official nodded to the rear of the lorry and tapped a clipboard with his pen.

Boris handed him a sheaf of papers. 'Explosives.'

The official stepped back. 'Who for?'

'Snow Science.' Boris pointed to the order. 'Want to take a look?' He jangled his keys.

'No, thanks.' The official held up his hands, palms outwards as if to shield himself from the cargo. 'What the hell does a research centre want with a lorryload of explosives?'

'The hell I know.' Boris shrugged. 'I'm just the delivery boy.'

'Better you than me, mate.' The official returned the papers and hit the button to raise the barrier. 'Drive on.'

Click. 46.5028, 13.7944. Intensity 152X, 648C

An hour later, the lorry crunched over rock salt as it swept up to the gates of Snow Science. In the amber dawn, the low buildings lay in darkness. Perfect. The only light shone from a Portakabin beside the main gate. Boris blocked the entrance with the lorry. He grabbed a hi-vis jacket and hauled a bag from the cab before climbing down.

The smell of smoked meat wafted through a gap in the guard-house delivery hatch. Something sizzling in a frying pan, bacon. Mmm. His mouth watered as he picked his way across the snow.

He tapped on the steamed-up window. 'Oi, Stefan!'

The gap in the hatch widened and the guard peered through.

'Delivery from Zagrovyl,' Boris said.

Stefan waggled a finger at Boris. 'Too early.' He pointed to the large clock behind him, a round white face with black roman numerals in a wooden frame. 'No one in. You'll have to wait.'

Boris swore and stamped the snow from his boots. 'You don't want to leave that stuff on the lorry.'

'Why? What have you got in there?'

'The usual.' Boris reached into the bag and produced a bottle of whisky. 'With the usual fee?'

Stefan stuck his head through the hatch and examined the gift before he accepted it. Strands of thin white hair blew around a freckled skull. 'It's cold out there,' he said. 'You'd best come in.' A lock clicked, and the door opened.

Thirty minutes later, the lorry rolled quietly into the Snow Science complex past several modern blocks separated by covered walkways. The special warehouse lay at the furthest corner of the site, behind a small hill. Boris knew where to go. He parked up and opened the lorry doors as Stefan arrived.

Stefan swept a torch over the contents and counted. He bit his lower lip while flicking through the papers on his clipboard. 'The order says two pallets.'

Boris handed a set of papers to Stefan. 'Here's the consignment note: twenty pallets.'

'That's ten times too much!' Stefan glanced behind him and dropped his voice. 'What do you want me to do, this time?'

'The usual.' Boris clapped him on the back. 'Call this number. Yuri will come and pick up the extra.'

Stefan shuffled from foot to foot. 'Why don't you leave it on the lorry?'

Boris shook his head. 'Regulations,' he said. 'Bloody European Union red tape and bureaucracy. They won't let me drive this stuff east without the proper paperwork.' He sighed. 'As soon as you call this number, they'll generate a new set of transport papers.'

Boris started to unstrap the cargo. 'C'mon. I've got other jobs to do.'

Stefan climbed into the cab of the electric forklift and manoeuvred it forward. After removing the first two pallets, he jumped down from the cab and sidled over to Boris. 'I'm not happy about this. Everything has changed here. Sergei has gone. There's a new boffin. A woman. She's more careful. I could lose my job.'

'How?' Boris rolled up the straps. 'When you're doing the right thing?' He unhooked a tarpaulin. 'You unload the delivery. Then you check it against the order.' He folded the tarp, once, twice. 'You notice something wrong,' he continued, doubling the sheet again: three times, four. 'You call the number I gave you.' He stowed the waxed canvas, neatly arranged into a square. 'Yuri picks it up. It'll be gone by the end of the day, before Dr Silver even gets notification of arrival.'

Stefan scratched his head. 'How do you know her name?'

Boris disappeared round the back of the lorry and reappeared, frowning. 'She placed the order,' he said. 'Dr Jaqueline Silver.'

Saturday 26 February, Kranjskabel, Slovenia

The halogen lamp flickered and then burst into light, illuminating the sign that straddled the entrance: SNOW SCIENCE. Set up as a collaboration between the European Space Agency and Earthwatch to compare satellite images of snow and ice with ground-based observation, Snow Science had expanded into a privately funded multinational, multidiscipline research institute.

The complex huddled high above the ski resort of Kranjskabel, hidden from view in the natural corrie at the top of a side valley, five kilometres of winding road from the centre of town, less than a kilometre as the crow flies.

Jaq took the direct route, sprinting up the mountain, snowboard jammed through the trolley sleeve of her bag-turned-backpack, breathing steadily. The spikes fastened to her running shoes skittered across tarmac then crunched over snow as she ran uphill, shortcutting the zigzags in the road. Long socks and padded overtrousers protected her ankles and shins from sharp crusts of ice broken by quick feet. The freezing air stabbed needles at her lungs; she exhaled in puffs of steam. A chill wind howled across the mountains as she emerged from the valley. Tugging at her woollen hat, she pulled the rim down over her earlobes and yanked the soft polyester snood up over her mouth and nose.

At the crest of the ridge she paused to admire a pair of buzzards. One bird cruised with the wind, a languorous tilt of black feathers fringing pale, broad wings to maintain course. The smaller raptor was working to impress, flying high before plummeting down in a crazy helix, twisting and turning on a roller coaster descent that made Jaq dizzy to watch. The plaintive *peea-ay* echoed between the mountains. Her heart soared as the male prepared to repeat his display, rising above snow-covered forests towards the saw-toothed peaks of the Julian Alps. A lone cross-country skier moved confidently across the horizon, the turquoise salopettes disappearing behind a cluster of pines.

Jaq retrieved her snowboard, clipped on and cruised down to her laboratory.

The perfect commute.

The gates of her workplace swung open and a lorry rolled out, the tail lights fading as it swept down the hill. An unfamiliar logo – Cyrillic script, Слив, SLYV – a Russian haulier driving all the way to Teesside and back for two pallets. How could that be good for the planet?

What was it like, the life of a long-distance lorry driver? Wages were low, drivers lived away from home for weeks, sleeping in the cramped confines of a cab. Now the most dangerous profession, the death rate on the roads was higher than any industry mortality, worse than mining. One day historians would revisit the twenty-first century and marvel at the abuse of the working man. *Logistics* sounded more benign than modern slavery.

Stefan waved from the guardhouse. Jaq released the clips and propped the snowboard under the hatch.

'I got your message.' She'd slipped away, while Karel was sleeping, to wash and change at her own flat. The run up the mountain had banished the dregs of a hangover. 'Am I late?'

'Come on in.' Stefan opened the door to his little cabin. 'I've got the delivery papers.'

She wrinkled her nose at the familiar fug of a single-skinned building with poor ventilation, eau de Portakabin with top notes from the drains.

Stefan pointed to the papers on the bench. Jaq squeezed herself past the row of CCTV monitors and found a seat. She perched on the edge, removed the bag from her back and wriggled out of her ski jacket. The stool tilted, and she had a sudden sharp memory of her school chemistry lab.

Her interest in explosions had started early. Even now she could evoke the smell of that first school lab: formaldehyde, vinegar and a faint whiff of leaking gas. It was a large room, south-facing. When the sun streamed through the windows, the green canvas blinds did little to impede the glare. Jaq and her fellow students sat on stools beside wooden benches, the mahogany surface scratched in places with the initials of bolder students.

The new chemistry teacher, Mr Peres, demonstrated the reaction of alkali metals with water, but used potassium instead of sodium. Rather a large piece. When the hydrogen ignited, with a beautiful lilac flame, he lost his eyebrows as well as his job.

Standing in the classroom surveying the shattered windows, the broken glassware and charred desk, Jaq had been truly impressed by the power of chemistry. It had provided her with gainful employment ever since.

Stefan coughed, and she turned her attention back to the paperwork on the bench. 'Two pallets delivered?'

Stefan flicked through the CCTV screens. 'In the quarantine yard.' He nodded at the screen.

Two pallets with danger labels plastered on the side stood in the snow, beside several dark squares on the ground.

'Did they deliver something else?'

Stefan turned away. 'They picked up some returns.'

The light was fading between the mountains as they walked to the warehouse. Keys jangled, tinkling in the cold, still air as Jaq pulled the bunch of twelve from the inside pocket of her bag. Stefan beat her to the first lock with his own, single key. As the outer door swung open, the lights clicked on and illuminated the delivery in the courtyard: two pallets, white bags with blue Zagrovyl labels. She crossed the open area, pulled off a glove and breathed on her hand before tackling the double lock of the inner door. The silver key with long shaft and serrated edge tugged reluctantly against the mortise lock, but the gold Yale key turned easily, triggering the timer.

Thirty seconds before the alarm went off.

Jaq strode through the inner door and approached the security panel. She ran her hands over the smooth plastic and pressed the top corners. With a snap, the cover dropped on its hinges, flapping under the bottom edge to reveal the flashing timer. Twenty-one seconds to go. She fumbled with the brass key – a fat hollow cylinder with wings, the shape of a clock winder. Finding the right-hand aperture, she pushed it in. Fifteen seconds to go. Now to get the sequence right. A quarter-turn left, a full turn right and half a turn left. The lights stopped flashing with eight seconds to spare.

High-energy explosives and detonators were kept behind blast-proof walls, but traditional propellants, like today's delivery, went to locked cages. She unlocked the empty cage with a four-sided key and signalled for Stefan to bring the pallets.

The warehouse sheeting creaked and quivered in the wind. A comforting rattle. The light construction was designed to flex and bend. Strength through adaptation rather than rigid resistance. In the event of an accident the blast would shoot upwards, blowing off the weaker roof panels. Unfortunate for birds, but safer for humans. The downside: it was freezing inside the store. The paperwork was complex, certificates of analysis to check against the batch numbers delivered, samples to take.

Stefan parked the forklift truck and got out, following her, stamping his feet and blowing on his hands, his nose and cheeks unnaturally red against a pale face and rheumy eyes. Poor man; he looked as if he could use a cup of tea.

'Thanks, Stefan.' She pointed over the artificial hill, in the direction of the gatehouse. 'Go and get warm.'

No point in both of them getting frostbite; she preferred to work alone.

Sampling. There was something pleasing about this most mundane of tasks. In her last job, she'd had a team to do this sort of thing, a team of people who needed managing; how refreshing to return to uncomplicated hands-on practical stuff.

The first pallet of explosives was smooth to the touch – tightly stretch-wrapped in clear film. Each one-tonne pallet had forty bags stacked in fours, ten high. Each bag contained twenty-five kilograms, the weight of a heavy suitcase. The standard acceptance protocol required sampling from 10 per cent of the individual bags, so four samples were required from each pallet.

Jaq selected the bags at random, marking a neat cross with a permanent marker. Snapping on a pair of disposable gloves, she cut a slit through the stretch-wrap. The sampling cylinder – a hollow metal quill with a sharp point – slid easily into the bag, freezing her fingers through the thin latex gloves. Swapping hands to turn the quill, she withdrew a sample and closed the perforation with special tape. The sample cascaded from the quill to a small glass bottle. She sealed it with a metal screw cap and added the date, time and the batch number to the label.

After taking four samples, she stretched and yawned. Not much sleep last night. A warm tingle lingered, embers of Karel's fire on her skin. She ambled over to the vending machine in the corner of the store and dropped a euro into the slot. Nothing happened. She slapped the side and pushed the coffee selection button again, but it remained obstinately silent. *Porra!*

The second pallet was in worse shape than the first, the stretch-wrap dirty and loose, and the bags had slipped in transit. She cut away the tattered film to inspect the delivery. Odd. The top bags appeared different from those below. The same plain white polyethylene, the same circular blue Zagrovyl symbol, the same label, the same hazard warnings, but the lower bags were misshapen and lumpy. Best to sample every one.

The top four bags were easy: the sample quill slipped in smoothly, the corer turned without resistance and a homogeneous column of white crystalline powder emerged. She poured each sample into a glass bottle and taped the centimetre-long perforation on the bag.

But the lumpy bags were problematic. The punch snagged and stuck and had to be tugged and wrenched out, the corer emerging with several different shades of powder along its length – white, cream, yellow, pink – in sharply defined layers. She emptied the sample quill carefully, labelling each bottle with the number on the side of the bag.

Despite the physical effort required to force the sample punch through the lumps, her teeth chattered. She stoppered the vial on the last sample and blew on her hands. Nearly done. One more task. Tick off the bag numbers against the delivery notes.

She jogged on the spot as she compared numbers. An icy finger of unease slowed her steps, freezing her with surprise. *O que é isto?* Jaq threw the clipboard and swore. She should have checked the numbers first. They didn't bloody match. The consignment note said one thing, and the bags said another. All that work, and she'd just spent hours sampling the wrong batch.

The windows rattled and the inner door flapped open. A gust of icy wind blew in a flurry of snow through the gap. High above the jagged peaks a new storm loomed.

She shouldered her bag, locked the warehouse and marched along the covered walkways to the office block. A row of brand-new snowsuits hung from a rail in the access hall, clear plastic covering flapping in the wind as she opened the inner door. The lockers rattled as she strode past, stuffed to the gunnels with climbing ropes, snowshoes, crampons, skis and poles.

Inside the laboratory, she placed the samples in the blast-proof fridge before dialling Zagrovyl complaints. Customer care, they called it. A misnomer if ever there was one. Closed for the weekend. Customer-don't-care, more like. She left a message. After locking up, she stowed the keys in the inside pocket of her bag and zipped it up.

Outside the snow was falling fast. She took a shortcut towards the exit gates.

'All finished?' Stefan opened the gatehouse hatch as she approached.

'Some mix-up,' she said. 'Consignment notes don't match the delivery. One of the pallets is quarantined until we sort it out.' She handed him the consignment notes through the hatch.

Stefan swayed and grabbed the window ledge. 'I'll call Zagrovyl.'

'It's fine,' she said. 'Already done.'

A strangled moan escaped his lips and a shadow passed over his face, descending like a shutter, leaving a sheen of sweat glistening on his brow.

Jaq stopped in her tracks. She'd sent him away earlier because of the cold, but now he looked seriously ill. A man in his sixties in a sedentary job with occasional bursts of physical activity in sub-zero temperatures. A recipe for disaster.

She pushed the handle of the gatehouse door. Locked. Heaving open the hatch, she yelled through the gap, 'Stefan, are you OK?'

He grabbed a slim tube from his inside pocket, opened his mouth and sprayed a short blast onto the back of his tongue.

A magical transformation took place before her eyes: a pink sunrise blooming over ashen skin.

Nitroglycerine. The liquid that killed one member of the Nobel family and inspired another to invent both dynamite and the Nobel Prize. Not just an explosive; also used to treat angina. A powerful vasodilator that relaxes the smooth muscles and opens the blood vessels to the heart.

Jaq stood rooted to the spot, her muscles suddenly frozen, holding her breath before exhaling in a long gasp. What a perverse symmetry. If she had known then what she knew now, could she have saved them? The men at Seal Sands. If they hadn't died, would she even be here now?

The hatch fell as she clenched her fists. No point looking back.

Lock it down. Lock it in. There was still time to help this man. She reached into her bag and found her phone. 'I'm calling 112.'

'No!' He opened the door.

Jaq took his arm and guided him to a chair. 'You need a doctor.' She checked his pulse – rapid but steady; his breathing – shallow but regular.

'I'm better now.' He grabbed her hand. 'You mustn't tell anyone else.'

She was taken aback by the fear in his eyes. Did this job mean so much to him? The lowest rung in the organisation, with the longest hours and the lowest pay. Snow Science was a good employer, regular medicals and adaptable work. They'd play fair with him. Stefan must be nearing retirement age anyway. Perhaps he was worried about his pension.

She sat with him for a while, making him a mug of hot chocolate, chatting about anything and nothing until he asked her to leave. Anxious to complete all his paperwork before the night shift arrived, he thrust the snowboard into her hands and practically pushed her out of the Portakabin.

New snow had fallen, leaving behind a thick, soft carpet which muffled all sound. As she bent to clip in for the descent, she glanced back at the lighted cabin.

Stefan stood at the open doorway, his back to her. One hand held a phone to his ear. He clenched his other hand into a fist and banged it against the wall.

Monday 28 February, Teesside, England

The limo driver held up a placard with FRANK GOOD, ZAGROVYL on it in large letters, and the Chariot Cars logo underneath. Frank's upper lip curled into a sneer. Teesside airport had fewer than ten flights a day. A clutch of chauffeurs loitered, waiting for the Amsterdam flight, the same uniformed drivers each time. The placard was hardly necessary – the limo driver knew Frank, and Frank knew his own name.

'Pleasant flight, sir?' The driver, PK, tucked the greeting sign under one arm and held out a hand for the luggage.

'Not particularly.' Tyche–Zagrovyl integration meetings were always tedious. Thank God that was the last one. Frank rolled the suitcase towards PK and strode for the exit. The castors clicked over the linoleum of the airport foyer.

PK caught the case and hurried after his client across the car park. 'Head office?' he asked as he opened the passenger door.

Frank nodded, his eyes never leaving the screen of his phone. Teesside airport might be convenient, but it wasn't worth a second glance: a single-storey building with a runway on one side and an access road on the other. Several pointless roundabouts joined the ambitiously sized airport car park – almost empty except for a patch sublet for caravan storage – to the A67 leading west to Darlington and east to Yarm.

The limo turned north. As they joined the A19, Frank snapped his fingers. 'Change of plan,' he said. 'We're picking someone up.'

Shelly stood outside her house, sheltering under a tree. Frank took in the high heels, silk blouse and linen skirt as the short raincoat blew open in the wind. PK jumped out and opened the door. She slipped in beside Frank, who barely acknowledged her greeting before turning his attention back to his telephone.

'Thanks, PK.' Shelly bent forward as the driver got back in. 'How are you?'

Frank scowled as the chatter between them interrupted his concentration. Shelly was a little too free with her attentions. Put him in mind of a poem he'd learned back at school.

A heart . . . too soon made glad,
Too easily impressed; she liked whate'er
She looked on, and her looks went everywhere.

Frank placed a hand on Shelly's knee. 'What time is the meeting?'

'Two thirty, Mr Good.'

'Then we have some time to play with.' He slipped his hand under her skirt, sliding it up to the top of her stocking where the texture became more interesting.

Shelly squirmed in her seat and glanced at PK.

'Don't worry about the driver,' Frank said. 'You know what to do, don't you, driver?'

PK met his eyes in the mirror. 'Transporter Bridge, sir?' The slow route to Seal Sands, crossing the River Tees on an Edwardian moving bridge. Frank nodded and pressed the privacy button. A dark glass partition rose into position, sealing the back from the front.

'Frank, please.' Shelly pulled his hand away. 'Not here.'

If not here, then where? Shelly had changed. Once upon a time, when he first hired her, he only had to lift an eyebrow and she'd start undressing. Frank shook his head and turned away. What was the point of her now?

The car drove past the ruins of Vulcan Street and down to the riverside. What a dismal picture. A giant blue structure towered above the river, a complex lattice of girders and rivets, carrying a suspended gondola designed to transport vehicles across without impeding the ships sailing up the River Tees. In the design office they called it an engineering marvel. It was more like an over-engineered dodo, a pointless waste of good steel. Made by men looking to the past, not the future. There were no tall-masted ships by the time the bridge was completed, and the new supertankers would never come this far upriver.

A plume of black smoke darkened the leaden sky and the smell of burning rubber wafted across the water. A steel ship that once operated as a nightclub lay tilted in the mud, abandoned and rotting. This was why change was needed. The derelict warehouses and empty wharves were the perfect reminder of how poorly those before him had performed. That's why Zagrovyl had chosen him as European operations director: a firm hand to put things right, or close them down forever.

Shelly leant against his shoulder, mouthing a faint apology, and began a laboured explanation. He stroked her long hair, letting his fingers meander down to her breasts. She sighed but didn't push him away this time as one hand slipped under silk. He knew what Shelly liked. A simulacrum of affection and she was his for the taking. Could he be bothered?

Under slanting rain, the platform of the Transporter Bridge sped across the ash-grey river to Port Clarence. When he realised she was crying, Frank jerked back, dabbing at his suit with a clean handkerchief before passing it to Shelly. The windows had misted up. He ran a finger across the glass and admired the glistening sheen before wiping it on her skirt, leaving a faint trace.

As the car neared the Seal Sands complex, Frank lifted the partition. Slumped against the door, Shelly stared out of the window. The tinted glass reflected a ravaged face: lipstick smeared across one cheek, black mascara smudged around enormous eyes. She looked fucked, even if he hadn't finished the job. But who was to know?

He pressed the privacy button and settled back as the tinted glass partition descended, linking his arms behind his head, and waited to make eye contact with the limo driver. How should a hireling respond? With a wink? Far too intimate; they were not of the same social standing. Acknowledge him with a nod? Inappropriate. Roll his eyes? A sure way to get sacked.

PK was no fool. He stared straight ahead.

'We're late, driver.'

PK glanced in the mirror. 'Yes, sir.'

The driver remained impassive, giving no hint of approval or disapproval, congratulation or censure. A disappointing outcome; Frank was spoiling for a fight. 'Drive straight to the warehouse.'

'Through security, sir?' PK bit his lip. 'I don't have a pass to enter the factory.'

Frank harrumphed. 'Shelly will sort it out.'

Shelly reapplied lipstick, smoothed her wrinkled skirt and checked the buttons on her blouse – a hasty attempt to repair the damage. At the gatehouse, she dashed out the moment the car stopped, without waiting for PK to open the door. Frank banged on the window and gestured impatiently at the security guard. The moment the barrier was lifted, Frank told PK to drive on, leaving Shelly behind.

The road ran through the centre of the factory. On the dockside stood the export cranes, rusted into dereliction. Opposite them the production units hissed and hummed, geometric sculptures of columns and spheres connected by a spaghetti of piping. At the end of the broad avenue stood the warehouse. When they arrived, PK jumped out and unfurled an umbrella before opening the passenger door.

Rain hammered on the roof of the limo. The wind swept the drops sideways, and they splashed back up from the pavement. Frank grimaced and sat back in his seat. He dialled another number.

'I asked for a report on controlled chemical stock movements.' Frank gestured for PK to close the door. 'I'm outside in the car. Bring it to me.' He scratched his crotch. 'Right now,' he added. 'I don't like waiting.'

Monday 28 February, Kranjskabel, Slovenia

Even Jaq had to admit that Zagrovyl responded quickly. Almost too quickly.

On Monday a man called her from the transport company. An error with some Zagrovyl deliveries, he explained. The pallet meant for Snow Science had gone to a warehouse nearby. The pallet they had delivered was reject material bound for disposal. He gave a long and detailed explanation. Almost too detailed.

He promised an immediate swap. The lorry was on its way. All very smooth. Almost too smooth.

The snow fell in slow, soft flakes, coating the Snow Science buildings in a fluffy white mantle, insulating the sophisticated laboratories from the primitive world outside. Jaq was catching up on some paperwork in her laboratory when the delivery lorry rolled through the main gate. She finished her report and locked it in a drawer before pulling on her jacket. The padded snowsuits lined the corridor, silent observers, empty limbs quivering as she made her way to the exit.

The security man swung the forklift round as she arrived, the replacement material already in the warehouse quarantine area.

'Hi, Patrice.' Jaq put up a hand, signalling for him to stop. 'Where's Stefan?'

'Day off. He's back on night shift tomorrow.'

The new pallet was in good shape, tightly stretch-wrapped and all the bags smooth and flat. Jaq unlocked the inner door, reset the alarm and opened the cage.

Patrice removed the rejected pallet and replaced it with the new one while she assembled her sampling equipment – pen, knife, gloves, quill and sample bottles. Boots crunched over snow. The inner warehouse door flew open. A bearded man, the delivery driver, stood in the doorway. Despite the cold, his unbuttoned tartan shirt revealed a chest as black and hairy as his beard.

He pointed at the bottles in her hand and frowned. 'Did you take samples from that other pallet?' Blackbeard inclined his head towards the lorry idling outside the door, now loaded up with a single pallet of reject material. He moved towards her, craning his neck to peer over her shoulder as she turned away.

Jaq locked the sample cupboard. 'It's okay, I'll dispose of them.' She carried the sampling equipment to the cage and shook out a pair of latex gloves.

'Give me the old samples.' The lorry driver advanced with an outstretched hand. He stood in front of her with his legs apart, chin jutting forward, a man who was not moving until he got what he wanted.

Jaq squared up to him. 'Why do you want them?'

'Reject material.' Blackbeard scowled, his thick brows meeting in the middle. 'Might get muddled up.'

Jaq brushed past him. 'I'll make sure it doesn't.' She crouched to check the bag numbers against the delivery note. This time they matched.

'Might be unstable.'

'We know how to handle explosives,' Jaq said.

'Look, lady.' He stamped a fur-lined boot. 'I was told to bring back all the material, and I do what I'm told.' He bent down so that his eyes were level with hers. Black eyes. 'So be a good girl and fetch them.' He reached out as if to pat her on the shoulder.

Jaq intercepted with the sharp end of the quill; it caught the side of his hand and he drew back with a cry.

At that moment Laurent sauntered in. Unusual for him to venture out of the office in such bad weather, but for once Jaq was glad to see her boss.

Blackbeard straightened up. 'Dr Visquel.' He shook Laurent's hand and introduced himself as Boris. 'I'm sorry about the mistake.

I was just explaining to Dr Silver that we need all samples back as well.'

'That shouldn't be a problem.' Laurent fixed his gaze on her. 'Should it, Jaq?'

How did Boris know their names? Jaq observed the way the two men stood facing her, Boris and Laurent. Close together, almost touching, a team. The blizzard had reached a new peak, the wind howling and snow falling so fast she could barely see the lorry through the open door, much less the laboratories and offices on the other side of the snowy mound. Icicles of unease chilled her spine; she shivered and shrugged away the apprehension. No point in arguing with these two.

'Let me finish up here,' Jaq said. 'Then I'll get the samples.'

'Anyone for coffee?' Laurent asked. He headed over to the vending machine. Jaq suppressed a smirk as it swallowed his coins. Laurent kicked it and tried again before inviting Boris to the office canteen.

'Bring them to the gatehouse,' Laurent instructed as he ushered Boris out.

Jaq collected four new samples and locked up. She chose the shortcut, scaling the artificial hill that acted as both helicopter landing circle and barrier between occupied buildings and explosives store. Bad decision. The snow had turned to hail. Buffeted by the wind, prills of ice lacerated her skin. She bent double and fought every step of the way before tumbling down the far side.

The office block offered sanctuary from the howling wind and stinging ice. She shook the snow from her hair, striding past the snowsuits that swung from a metal rack in front of the lockers. Inside the lab, she scanned the room. Rita, the analyst, sat in the far corner, engrossed in a phone call. All clear. Jaq took a deep breath before selecting four of the forty samples, the ones from the top bags.

Jaq shoved the samples into her jacket pocket and headed back out into the blizzard. The gatehouse lay opposite the car park, but the wind whirled the fallen snow into vortices of pure white-out. She could barely see her hand in front of her face. She wasn't going to make the same mistake twice. Better to take the long way round, through the walkways. It meant retracing her steps, but the partial cover offered some protection from the storm. Jaq lowered her head and battled on to the gatehouse. Blackbeard jumped down from the cab of his lorry.

'Here you go.' Jaq handed over the samples.

He scrutinised her face. 'This the lot, then?'

'We sample one in ten.' She maintained eye contact. 'Standard procedure,' she added before stepping back into the storm.

Monday 28 February, Teesside, England

Frank strode into his office and heaved open a window. Outside, the ancient factory sprawled towards the River Tees. His eyes followed the progress of a Russian ship as he loosened his tie. Insufferably hot again. He'd sack whoever kept the offices so warm. A waste of money, stifling thought and creativity, curbing action. Only the elderly and the lazy needed inside temperatures above 18 degrees Celsius. He'd have a word with the engineers about moving the control panel into his office and locking the damn thing.

Robin put his head round the door. Tufts of brown hair speckled with grey framed a pale, bespectacled face. His brown, beady eyes scanned the room in quick, jerky movements. Dressed in a brown suit with white shirt and red tie, the finance director looked more birdlike than ever. 'We're all assembled and ready when you are,' he said.

Frank waved the bean counter away. He removed his outdoor coat and draped it over a wooden hanger, smoothing the tan cashmere before suspending the hanger from a curlicue on the hat stand. He paused for a second to admire the photo on the wall. It showed Frank – in white shorts, navy shirt and captain's hat – taking possession of his new yacht, purchased with the bonus awarded after the Tyche acquisition. *Good Ship Frankium* was waiting for him in Cannes. The sooner he concluded this business, the sooner he could get back to the Med.

Double doors separated his office from the boardroom. He flung them open, and a hush fell over the management team. The newer managers, the ones he had hand-picked, sat bolt upright, jackets off and shirtsleeves rolled up, expectant, eager. The dead wood, the managers he had inherited, were slumped in their seats, jumpers on under jackets, stinking of fear.

'What's first on the agenda?' Frank asked, grabbing the papers laid out on the table.

'Safety,' Robin said. 'Stuart, do you have anything for us?'

Frank drummed his fingers as the safety manager fiddled with the projector.

'Transport of hazardous goods regulations,' Stuart said. 'I'd like to update the team on how the EU harmonisation plans will affect us in Eastern Europe.'

Frank yawned. He allowed Robin to run the meeting, feigning interest as the department heads churned through their updates, his fingers tapping on the polished mahogany table, running through the first Brandenburg Concerto in his head. On and on they droned, spinning the good news, glossing over any difficulties, concocting elaborate excuses for production targets missed, sales contracts not signed, budgets overspent: the usual crap. Frank was just getting to the Minuet when Shelly tiptoed into the room, clutching some papers.

Look what the cat dragged in, Christ Almighty, but she looked a mess: hunched, crumpled, squalid. How could he ever have dipped his pen in that ink?

He signed the papers and waved her away. The meeting dragged on, but he'd lost his train of music.

'Any other business?' Robin asked, scanning the table.

Frank could see the group relax, amazed that a meeting with him could have turned out so tranquil, so amicable, so civilised. So boring and unproductive – time to pounce.

At the other end of the table, the HR director raised a hand. The ugly dwarf had no pretentions to glamour. Nicola was sensible shoes and supermarket value-pack knickers; he could tell without ever having seen them.

'About the team-building event—'

'Cancel it.' Before Nicola could protest, he held up a hand and continued. 'We have one other item to discuss.'

He strode to the full-length window. His mouth hardened as he stared out at the river. Four o'clock in the afternoon, and it was almost dark outside. The sodium lights cast an egg-yolk glaze over the towers and open staircases of the production buildings. Steam puffed into the sky and the intermittent screech and bang of conveyors carried through the pitter-patter of rain. God, it was an ugly shithole. Built in the 1980s and falling to bits. Well, its days were numbered.

'I'm cancelling the UK expansion project,' he announced.

The distributed murmur rose to a crescendo of confused protest. Frank addressed the window, admiring his reflection, speaking softly to force them to stop whining and listen. 'We are competing with the giants of the developing world – Brazil, Russia, India and China. That's where our future lies. Those are the only expansion projects that will get funding in future.'

'Projects like Smolensk Two?' asked Eric, the dry Scottish voice of the engineering manager dripping with venom.

'Exactly.'

'Late, incomplete and already over budget?'

Frank shot him a look of contempt. 'Smolensk Two is already in production.' Wasn't it? Suddenly uncertain, Frank stomped back to the table and leafed to the page on the board papers with production figures. 'Page seven,' he snarled. 'Or can't engineers read?'

Robin shook his head. 'That's not new production. It's recycling. Rejected product collected from customers in Europe and sent to Russia for rebagging.'

Frank stood still, erect, alert. 'Explain?'

'The labour costs here are too high to make recycling economical. So, reject goes east for recovery,' Robin said. 'And actually, there is a problem with the Smolensk numbers—'

Frank interrupted him and addressed Eric. 'When was the Russian expansion due to start up?'

'Last quarter.'

'And?'

'Problems with equipment delivery, I believe, but—'

'Why did no one tell me?'

Robin and Eric exchanged glances. 'The UK engineering team were . . . not involved with the Smolensk project.' Robin chose his words carefully.

Eric was more forthright. 'You specifically excluded us, told us to keep our bloody noses out and leave it to Ivan. Told us we slowed projects down by insisting on proper engineering studies and—'

'Intolerable excuses.' Frank slammed his fist onto the table. Executive reward was heavily weighted towards international expansion. If any target was missed, then his bonus payment would vanish. Russia had to produce. How else was he going to keep his yacht? 'I want a full report on my desk by tomorrow.'

'Then you'd better call Ivan yourself,' Eric retorted.

'Get out!' Frank shouted. 'All of you.'

The team didn't need to be told twice.

Frank glared at Nicola as she waddled round the table. Why was she always last? She wasn't his appointment, but it was easier to leave her in post until he'd finished pruning the team.

It sometimes amused him the way the fat cow tried to conceal her animosity. She didn't bother to smile any more, but nor did she bare her teeth and hiss, which is what he suspected she wanted to do right now.

Today he was not amused.

'Nicola, wait.'

He stared down at her. She must have remarkably short legs. 'Could you have a word with Shelly about her appearance?' he said. 'I think it would be better coming from another woman.' He smiled internally as she flinched. 'I know it has been difficult for her since the bereavement,' he continued, 'but I expect certain standards to be upheld. She was looking positively bedraggled today.'

'I'll deal with it.' Nicola nodded and started to turn away. Not so fast.

'That new girl, Raquel, can stand in as my PA for the next few trips.' Frank paused, choosing his next words carefully. 'Until the situation is resolved.'

Nicola wheeled round and stared directly at him. Her mouth opened and closed like a fat, wet goldfish, eyes darting left and right. Was she considering her options? He returned the eye contact, staring into her soul, daring her to fight back. She lowered her eyes. Lily-livered lackey, she was cunning enough to pick her battles. There was no love lost between ugly Nicola and once-stylish Shelly. The HR director took a deep breath and spoke slowly, choosing her words carefully. 'I understand and share your concerns about Shelly,' she said. 'Leave it with me.'

The repulsive little troglodyte had a special vacant expression that annoyed him much more than open rebellion. The lights were on, but no one was home. Nicola was not stupid; she had locked away part of her spirit. He saw that she was not afraid of him. Perhaps he could change that.

Frank returned to the window. A Russian ship approached the dock, a mournful honk answered by a sharp toot from the tug boat guiding her to berth.

The Smolensk production expansion was late. Why had no one told him? Ivan and his team needed shaking up. Time to go to Russia and do it himself.

Tuesday 1 March, Kranjskabel, Slovenia

Jaq arrived before dawn to prepare the explosives for the helicopter crew. Routine stuff, create a shock wave just above the snowpack. No experiments, no data collection – this was the kind of job the air crew liked best. Hurling explosives out of a moving helicopter, wireless detonation and suddenly a cascade of snow came spilling down the mountainside and it was safe to ski again. She often went with them for the ride, but today she had other things to do. Thirty-six suspicious samples to analyse.

What was in the reject pallet? Jaq had worked with explosives long enough to know the material in the lumpy bags was not pure ammonium nitrate. Crystalline powder, yes. Pure white, no. The colour was not conclusive; it could come from impurities – traces of iron gave a pink tinge, heat damage a yellowish hue. But there was something else about the samples that had nagged her all evening. Something fishy. One bag smelt fishy. Literally fishy. The stench from rotting fish. Or certain chemical compounds. Just a whiff, no more.

Why had the lorry driver been so anxious to get the Zagrovyl samples back? Why had Laurent been so keen to comply? And why had he behaved so strangely?

After the lorry left with the reject pallet, her boss found reasons to keep her from the lab. Laurent never held impromptu meetings – a slave to his calendar – and yet he suddenly insisted they meet and talk about some insanely dull improvement programme he was launching. Then there was a scheduled meeting about today's blasting, after which Laurent escorted her back to the laboratory. She jumped at the chance to escape from him when Rita offered her a lift home.

Laurent, like most bad bosses, hated to be challenged. So she deferred her plan to examine the samples again. Because she had to book out the explosives at daybreak anyway. And Laurent was not an early riser.

The stars faded as Jaq locked up the warehouse, the opaque sky-ink bleeding from black into translucent blue. She saluted the helicopter as it banked overhead, her heartbeat accelerating with the whirring blades. On the far side of the snowy hill the laboratories awaited her, square white rooms with grey benches, stuffed with the analytical tools to unlock any mystery. Time for action.

Chemistry had moved on from the days of the school lab. Now the benches were crowded with machines, featureless boxes of varying shapes and sizes, all connected to computers.

The eight normal samples – four from the first pallet and four from the replacement pallet – had already been processed by the lab. Jaq checked the results. All good. Approved.

The Italian analyst, Rita, arrived and bid Jaq a cheerful good morning, adding, 'Can I help?'

'Thanks.' Jaq made a decision. 'But this is one I need to do myself.'

Jaq donned a white coat and safety glasses before opening the fridge. She placed the thirty-six little glass bottles on a steel tray and lined them up in six rows of six, labels facing forward. The preparation was simple. Gloves on. Unscrew the cap. Use a thin metal spatula to remove a few milligrams of powder. Tip it onto a transparent quartz disc about the diameter of a two-pound coin but thicker. Press another disc over it and make a sandwich. Slot into the carrier. Press a button and see it disappear into the black box. Replace screw cap on sample bottle. Remove gloves. Note down sample number and time. Repeat thirty-five times. Plus one – a sample of pure ammonium nitrate, the standard, for calibration and comparison.

As the results spilled out, she scratched her head. Just as she thought. Not ammonium nitrate. But what was it? More tests required.

Her phone pinged. A text message.

Can we meet?

It was signed by someone she had never heard of, *Dr C. Hatton.*

Wrong number. She ignored it and programmed the next tests.

Ping. Another message appeared.

From Zagrovyl.

The spatula clattered to the floor. *Caramba!* Were those bastards telepathic? She squinted at the halogen lights above the bench. Had Laurent installed cameras up there, or was information being streamed from the analytical machines straight to Zagrovyl? Was the mysterious Dr Hatton watching her as she puzzled over the results of the samples she wasn't meant to have? She made a face, sticking out her tongue and rolling her eyes for the invisible camera.

'Everything okay?' Rita retrieved the spatula and placed it back on the tray, suppressing a smile.

Jaq hung her head, a flush of embarrassment warming her throat. Get a grip. She dialled the number for Dr Hatton. It went straight to a generic voicemail. She didn't leave a message.

Ping! Jaq jumped.

Not phone. In person. Urgent. Café Charlie. 11am?

Why did someone from Zagrovyl want to meet with her? And why in a café in the centre of Kranjskabel? Why not here? Well, they could whistle for it. Book an appointment via Sheila, the department secretary, like any normal supplier. She got back to work.

Rita hovered. 'Are you ready for the inspection?'

'What inspection?'

'Dr Visquel. He's doing random 5S checks at eleven.' Rita coughed. 'Best to tidy everything away.'

Santos. Laurent and 5S. Another of his gobbledegook management initiatives. Best avoided.

Jaq checked her watch. Ten thirty. She pulled back the blind. The storm had finally blown itself out and the sky glowed bright blue again. She could ride down. Clear her head. Leave the samples on to run against a range of different standards. Try to make some sense of the results when she returned.

Jaq grabbed her bag. 'I need to go out for a bit.'

Rita grinned. 'Good idea.'

First tracks. The snowboard cruised over the pristine surface, powder over hardpack, smooth on the slide with enough bite for the sharpest cuts. Crunchy.

Jaq made it to the edge of town before swapping the snowboard for spikes. As she bent to unclip, a screech reverberated up a narrow side street. It took her a moment to identify the source. The red-and-white-striped awning of Skipass restaurant was unravelling, lumps of snow falling from the folds as it stretched.

A snowplough rumbled past, followed by a stream of cars. A welcome smell of coffee hit her as she entered the brightly lit café.

Jaq stashed her board in the rack and scanned the clientele. Skiers. A group of teenagers. No one who resembled a chemical company representative. She headed towards an empty table when a woman in turquoise salopettes waved from a booth at the back of the café.

Jaq approached her. 'Dr Hatton?' she asked. 'From Zagrovyl?'

The woman nodded. 'Camilla,' she said, and extended a hand with painted nails.

Jaq removed a glove, her hand cold in contrast to Camilla's warm handshake. Like Jaq, she wore no make-up and the uneven tan suggested she spent time on the high slopes, wore goggles and took the sport seriously. Aged anywhere between forty and sixty, impossible to tell; expertly styled short, white hair and startlingly green eyes.

'May I call you Jaqueline?' Her English had a hint of Central Europe, or perhaps Scandinavia.

'Jaq is fine.'

'I wasn't sure you'd come.'

Quite right. What on earth was she doing here? 'I needed a decent coffee.' Jaq ordered an espresso from a passing waiter, and Camilla asked for more water.

'Here, let me help you.' Camilla seized Jaq's bag and hung it on a hook on the edge of the booth, draping her ski jacket over it. 'So, Jaq, how long have you been with Snow Science?'

The question was friendly, an icebreaker, but the eyes were searchlights sweeping over Jaq's face, not the eyes of someone who indulged in small talk. So Jaq did not indulge her.

'I assume you know about the mix-up?'

Camilla dropped her eyelids, hooding her eyes. 'Some delivery problems? All sorted now?'

'I took samples.'

Camilla met her eyes and blinked, reassessing. 'And you analysed them.' She nodded to herself as if she would expect nothing less. 'So, what did you find?'

Jaq scanned the room. This was not an appropriate place for the discussion. 'Why did you ask to meet me here?'

'I'm on holiday, so this is more of a courtesy call.' Camilla smiled a radiant blast of charm. 'I thought maybe we could keep this informal – talk off the record?'

'Far too serious for that.' Jaq made to stand up.

'Why?' Camilla put out a hand to detain her. 'Why are you worried about reject material scheduled for recycling?'

It was a good question. Why was she wasting time on it? Because everyone seemed so determined she shouldn't. Not good enough. That just sounded contrary. Where to start? Because it smelt fishy. That sounded mad. 'Because something is being transported in wrongly labelled bags.' That sounded lame.

Camilla's eyes narrowed. 'What do you mean?' She didn't seem entirely surprised, more like someone feigning surprise and doing it badly. Dissembling. Real surprise makes your eyes open wide.

Jaq reached for her bag. 'Come with me to Snow Science and I'll show you.'

'I can't,' Camilla said. 'Your boss is Laurent Visquel?'

Jaq paused. 'Yes. What about him?'

'We don't get on,' Camilla said.

You and me both. The waiter arrived with coffee and water. Jaq sat down again. 'Why?'

'It's complicated.'

Jaq shook the packet of sugar by the corner. She tore off the top and tapped half the contents onto a spoon above her cup. 'Come on, you'll have to do better than that.'

A man with a briefcase brushed past the booth; Camilla waited until he was out of earshot. 'You remember Laurent's paper on artificial glaciers?' she asked.

Snow Science received massive funding to research alternatives for water storage. Over half the earth's fresh water is stored in glaciers. And glaciers are melting. Jaq nodded and tipped the sugar from her spoon into the coffee.

'There is a civil engineer in Ladakh who spent his retirement creating artificial glaciers. Laurent stole his data.'

Jaq stirred her coffee. It didn't surprise her. Her boss was lazy. And prickly. 'Laurent wouldn't like being found out.'

'It wasn't just the plagiarism.' Camilla frowned. 'Laurent made some dangerous simplifications. He totally underestimated the risk of flash floods.'

Jaq nodded. She'd worked on the use of explosives for controlled release from glacial lakes. 'What happened? Did you make this public?'

'No.' Camilla shook her head. 'A private agreement. Laurent retracted the paper and reissued it with acknowledgements and corrections.' Camilla smiled. 'And the Ladakhis received a generous donation towards their research from the private purse of an unknown benefactor.'

Jaq laughed, suddenly better disposed towards Camilla. *My boss's enemy is my friend.* She sipped her coffee.

'Your publications, on the other hand, are beyond reproach,' Camilla said. 'I particularly enjoyed your paper, *Natural Danger*. People forget about natural poisons like ergotamine and botulism.'

Jaq acknowledged the compliment with a little flash of gratification. A reminder of happy days as a visiting lecturer at Teesside University; the paper had caused some controversy at the time but was now largely forgotten. Either Camilla had seen it when it was first published, or she had done some thorough research. A shiver of unease rippled across Jaq's skin. What else did she know about? The inquiry that had almost ended Jaq's career, forced her to seek this quiet research job far from Teesside?

'What is your role in Zagrovyl, exactly?'

Camilla reached for an inside pocket and fished out a business card.

Camilla Hatton, Zagrovyl, Director of Change.

'A roving remit. I'm a troubleshooter.'

'Well, you have some trouble to shoot at here.' Jaq inspected the face of the woman opposite. Intriguing. She didn't fit the traditional mould of a Zagrovyl director. Female, that was rare enough, but – unless Jaq was completely mistaken – a core of integrity shone through. Why else would she take an interest in Laurent Visquel's perfidy? Or was that a smokescreen? She was hiding something, knew more about the mislabelled bags than she admitted. The address on the card signalled the Teesside office, head office. A place Jaq knew only too well. Someone in Zagrovyl must be worried if they were sending a director from head office to Slovenia.

Camilla glanced over her shoulder and then leant forward, lowering her voice. 'Jaq, do you know where Sergei is?' There was an unusual quality about Camilla's voice, a grave intensity.

Jaq raised an eyebrow. 'Who the hell is Sergei?'

Camilla put a finger to her lips and whispered. 'Sergei Koval, your predecessor at Snow Science.'

Jaq had a vague memory of hearing the name. 'The Russian?'

'Ukrainian.'

Jaq shrugged. 'I haven't a clue.'

'Has he made contact with you?'

'No, why should—'

'Did he leave anything for me? A key?'

A twist of unease spiralled from Jaq's stomach and tightened her throat. A key? Why would a Snow Science employee leave a key for a supplier? 'No.'

Camilla glanced over her shoulder again. 'Notes? Files? Data? A memory stick? Anything? You sure he didn't leave a key?'

Jaq tried to follow her gaze. Nothing but skiers and holiday-makers crowded at the coffee bar, paying no attention to two women in a corner booth. Why all this fuss and drama? She raised her voice, establishing her refusal to engage. 'Why would he leave you a key?'

A shaft of sunlight penetrated the back skylight, making Camilla's white hair glow. Ethereal, angelic. 'I have to find it,' she whispered.

Jaq pressed against the padded seat, distancing herself. 'I don't understand,' she said. 'What is your connection with Sergei Koval?'

Camilla shifted in her seat as the waiter approached. She waved him away and checked the room before continuing. 'He contacted me about some irregularities.'

'Irregularities?'

'Some discrepancies with Zagrovyl-sourced products.'

'What sort of discrepancies?'

'I wish I could tell you more.' Camilla sighed. 'Before we could meet, he vanished.'

Vanished. Jaq wrapped her arms around her chest, suddenly cold.

'Why did Sergei contact *you*?' Jaq said. 'Don't Zagrovyl have special departments to investigate quality complaints?'

Camilla pursed her lips and wrinkled her nose. 'Sergei knew that I was investigating something a little wider.'

'Involving Zagrovyl, your own company?'

'Involving the whole supply chain.'

A prickling sensation on her skin made Jaq want to scratch, but she resisted the urge. 'So, you check how Zagrovyl chemicals are used after they leave the factory?'

'Something like that,' Camilla said.

'And Sergei had information?'

Camilla nodded.

'But he left before he could give it to you?'

Camilla scrutinised her face. 'He told me he was keeping the evidence in a safe place, and when we met he would give me the key.'

'As in code, password?'

Camilla raised her glass to the light, watching the colour change as it rotated. She appeared to come to a decision. 'I took it to mean a real physical object.'

'Camilla, what exactly do you want?'

'I'll level with you,' Camilla said. 'It's personal, not official Zagrovyl business. But it *is* important.' She fixed Jaq with a penetrating stare. There was pain there, and passion. 'Sergei wouldn't communicate by phone or email. He wouldn't talk to anyone except me, and then only face to face.' She lowered her head. 'I will never forgive myself if anything has happened to him.'

Jaq's stomach did a double somersault. *Happened to him?* Sergei had moved on. That meant he'd had enough. Packed it in. Resigned. Found something new. 'Why should something have happened to him?'

Camilla dropped her gaze and inspected her manicured finger-nails. 'When I heard that you had taken his place, I wanted to meet you,' she said. 'In the hope that you might help me. But also to warn you.'

Jaq frowned. 'About what?'

'Jaq, we need to find that key.'

Who *was* this woman? Since when did *she* call the shots? 'There is no *we* here, Camilla.'

'Then *you* need to find the key,' Camilla said.

Jaq gave an exasperated snort. '*I* don't need to do anything.'

Camilla placed both her hands on the table. 'Come and work for me.'

Work at Zagrovyl again? No, thank you. Once bitten, twice shy. 'You must be joking.'

'I'm deadly serious,' Camilla said. 'Think about it.'

Jaq sprang to her feet. 'I don't need to think about it. The answer is no.'

'Be careful, Jaq. The less you know, the safer you'll be.'

A threat?

'This is part of something big.' Camilla dropped her voice to a whisper. 'Strange things may happen. Keep away. Stay safe. Don't get involved. It could be dangerous.'

Basta. Enough. Jaq seized her ski jacket. The hook underneath was empty, bare, shining brass. Where was her bag? There, on the floor, just below the hook. How the hell had that happened? She reached down to retrieve it, her hand darting to the inside pocket. She muffled a gasp of relief as her fingers curled round the bulge of keys. All secure.

Camilla bent down and caught Jaq's arm as she rose, holding on longer than was comfortable. 'If you won't help me, stay out of this, Jaq,' Camilla said. 'For your own safety.'

Rushing to retrieve her snowboard, Jaq almost collided with a man backing into the café, shaking his snowy boots onto the street.

'Jaq!'

Surprised? Certainly. Dismayed? Impossible to tell. At least he remembered her name.

'Hi, Karel.'

'It's great to see you,' he said.

A throwaway phrase. Polite and meaningless.

He kissed her on both cheeks with cool lips. 'Can I buy you a coffee?'

'I'm just leaving.'

He tilted his head, blue eyes flashing under fair lashes over cheekbones like razors.

'Another time?'

'Sure.'

'I wanted to get in touch.' His cheeks flushed. Did he blush when he lied? Or was she being unfair? The effect of the warm café after the winter air outside. 'I don't have your number.'

Jaq hesitated. What did it matter? She reached into her jacket and handed him a business card. A movement from a booth caught her eye. Camilla was standing, peering across at Karel. Time to leave. A bus inched up the hill. The shuttle for Snow Science. Faster than running.

'Must dash.' Jaq made a phone sign with her thumb and pinkie. 'Call me?'

The bus stank of sweaty socks. She nabbed a seat by the window and rubbed condensation from the glass as the bus trundled past the café. No sign of Camilla following her. Good.

Through the café window, she glimpsed a flash of golden hair as Karel bent his head to talk to someone. She couldn't see who it was, nor did she really care.

Jaq had some chemical detective work to do.

Tuesday 1 March, Kranjskabel, Slovenia

By the time Jaq returned to Snow Science, the samples had vanished.

She stood in the centre of laboratory number five and blinked. *Que loucura!* No tray, no spatula, no samples, no quartz cells, no printouts. The bench tops blinked back at her, the bare surfaces polished and shining. Everything gone. But where?

The low hum of the machines mocked her as she searched in drawers and cupboards. Nothing. She checked the device that had been rerunning the samples while she was out. It had powered down. She opened it up to retrieve the sample cells. Empty. She called up the memory to see the results of the last runs. Wiped clear.

Little bubbles of rage expanded as they rose, exploding with a howl of fury. No one heard her. Laboratory Five was empty. Not just of her samples and results, but of people as well. No Rita. What was going on? Jaq ran, searching for someone to question. A flash of silver from the snowsuits, the reflective strips sparkling under the corridor lights; limp, headless puppets mocking her. No one in sight. She followed the corridor down to the training block and ran right into a crowded lecture theatre.

'Ah, Dr Silver, so glad you could join us!' Laurent stood on the podium in front of a crowded auditorium. Jaq groaned inwardly. She had totally forgotten about the training, the new initiative launched by Laurent – Workplace Organisation and Standardisation. 5S. Everything in its correct place. Tidiness for idiots.

'Take a seat.' Laurent made it a command, not an invitation, indicating an empty row at the front.

Jaq ignored his accusatory finger and slipped into a back row next to Rita, the analyst from Laboratory Five.

'As I was saying,' Laurent resumed the lecture, 'the Japanese first introduced 5S. It translates as "sort", "set in order", "shine", "standardise" and "sustain".'

'What happened to my samples?' Jaq whispered to Rita.

Rita frowned and shrugged.

'They were on the bench, next to where you work,' Jaq hissed.

'I thought you'd finished,' Rita whispered back. 'All tidied away. His orders.' She nodded her head at Laurent, who was explaining the benefits of the new workflow system.

'Where to?'

Rita mimed a dumping motion.

'The bin?' Jaq hadn't thought to check the bin. There was still hope.

Rita shook her head and whispered, 'Recycling.'

The incinerator. 'What the fuck!' The words exploded and ricocheted around the lecture theatre. Several heads turned.

Laurent ignored the interruption, adopting his most lugubrious smile as he welcomed several colleagues to the podium to explain the detailed roll-out of 5S to Snow Science.

There might still be time. The samples were small, a few grams from each bag. The Snow Science incinerator was designed to handle explosives. But everything was checked first. Jaq sprang to her feet and exited the training, letting the door bang behind her.

She sprinted down corridors, cutting across the snowy mound that separated the warehouses from the laboratories and offices, swinging left across the helicopter landing circle and into the utilities section. Wisps of steam rose from drains in the boiler house, a constant *psst! psst!* of steam traps discharging condensate into the tiled drain channels. She ran past the hiss of a pneumatic air leak and the roar of a diesel turbine. Faster. She was breathing heavily by the time she arrived at the recycling section.

Too late.

The operator opened the sight glass on the incinerator and Jaq bit her lip as the samples fell into the flames.

'Dr Visquel said to give this priority,' the operator said.

Oh, he did, did he? The underhand, treacherous, useless snake.

The glass melted, the powder sizzled and flared with a series of exploding blue and orange fireballs.

Jaq waited for Laurent outside the lecture theatre.

'Dr Visquel, we need to talk.'

'Yes,' he said. 'I think we do.' He gestured towards his office.

Sheila, Laurent's secretary, turned as they entered, greeting Jaq with a question about the warehouse vending machine. 'Small triangular key with two prongs. Can't find it anywhere. Have you seen it?' Jaq raised her eyebrows to warn her that now was not the time. Sheila winked behind Laurent's back, locked the filing cabinet and left, closing the connecting door.

'Why did you destroy my samples?' Jaq kept her hands clasped together, focusing her fury on keeping them still.

Laurent sauntered over to his desk and sat on a leather chair.

'I've no idea what you're talking about.' He opened a desk drawer and extracted a pair of nail clippers. 'And before we discuss anything else, how do you explain your outburst in my lecture?'

'I was working in Lab Five.' Jaq laced her fingers together and squeezed, dissipating the anger so she could keep her voice level and slow. 'I was analysing some samples and they disappeared.'

'Take it up with the lab technicians.'

Typical. Find someone else to take the rap. 'I already spoke with Rita. She said you ordered it.'

Laurent inspected the nails on his left hand.

'If you were absent,' clip, clip, 'for the 5S inspection,' clip, clip, 'then you have only yourself to blame.' Clip. Laurent snipped the final nail and swapped the clipper to his other hand. 'Did you book the experiment into the logbook and get an SP5 sample number?'

The man was insufferable. 'There was no need. I was doing the analysis myself.'

'Did you label the samples with the SB5 barcode?' Clip. Clip.

Worse than insufferable, a halfwit. 'I already told you . . . I was trying to identify them.'

'Did you program the machines with the SP5 sample number?' Clip. 'Or scan in the SB5 barcode?' Clip.

No, not a halfwit, an evil orc. 'You know I didn't.'

'Did you write up your results on the SR5 results sheet?' Clip.

Jaq clenched one fist and slammed it into the palm of her other hand. 'Stop pretending this is about lab protocol. You destroyed my samples deliberately. You gave it priority. Why?'

Laurent swept his hand across the table. The nail clippers jumped and clattered to the floor. 'I will not have the programme jeopardised!' Little flecks of spittle dusted his moustache. 'We get the quality we are prepared to accept. Zero tolerance to deviation is the only way to raise standards. I will enforce it. Do you realise how important this is for the future of Snow Science?'

'What, 5S?' Jaq laughed out loud. 'Glorified housekeeping?' Could he be that stupid? No, he was hiding something. 'I don't believe you.'

Laurent bent down to retrieve the clippers. 'I gave you every opportunity, Jaq,' he said. 'I tried to help you to integrate.' He shook his head slowly. 'But you spurned me.'

Christ, was that what it was all about? Rejecting Laurent's clumsy pass in the karaoke bar? Leaving with a gorgeous young stranger instead of shagging the ugly old boss? Good lord, what century did Laurent think he was living in?

Laurent folded his arms across his chest. 'The funding from the consortium depends on us introducing normalised working practices.'

Oh, to wipe the condescending smile from his face with her boot. 'It is not normal to destroy other people's samples,' Jaq said.

'We have systems,' Laurent said. 'We have protocols. If you

are not with us, then you are against us. I can't have one arrogant individual ruining the efforts of the team.'

Arrogant! That was rich, coming from him. 'In what way am I—?'

'Enough!' Laurent held up both hands, palms outwards. 'I am willing to forgive your profanity in the lecture theatre, your wild accusations and disrespect in my office, your failure to observe laboratory protocol, but only,' he slammed his hands down on the desk, 'if this is the last time I hear about those blasted Zagrovyl samples. Do I make myself clear?'

Oh yes, perfectly clear. Laurent was taking instructions from Zagrovyl.

She slammed the office door as she left.

Tuesday 1 March, Jesenice, Slovenia

Boris secured the load and closed the curtain of the SLYV lorry. Loaded with eighteen pallets disguised as explosives, the SLYV wagon was finally ready to roll.

He chanted along to *Masters of Rock* on the approach to Lake Bled, the rising sun blinding him as he turned east. Sliding on a pair of sunglasses, he paused to stare at the fairy-tale castle perched on a cliff above the frozen water and frosty island. General Tito once kept a summer house here. Good choice. They didn't make dictators like they used to.

Boris bypassed Ljubljana and crossed the border into Hungary at Pince.

Click. 46.53320, 15.60110. Intensity 72*X*, 648*C*

A customs officer waved down the lorry and indicated a side bay. Boris pressed pause on 'Time to Change'. The thrum of the engine replaced Pavel Slíva's guitar riff as he rolled down the window.

'Where did you pick this up?'

Boris handed over the paperwork. 'Snow Science, Slovenia.' He watched in the side mirror as she took the papers and walked to the back of the lorry, his mouth suddenly dry. Had he forgotten anything? Yuri normally made this run. But Yuri had fucked up once too often. Picking up the wrong pallet, leaving one behind that could expose the whole supply chain. Blow the operation wide open. Boris had taught Yuri a lesson. Yuri wouldn't be driving again. Not for SLYV, and not for anyone else.

She was back. 'Final destination?'

'Smolensk.' Boris pointed to the false delivery address on the paperwork. 'Zagrovyl, Russia.'

'Open up, please.'

'Back or side?'

'Everything.'

Was she on the make? Should he have slipped her some cash with the documents? Her body language didn't invite it. Sour-faced *knedlík*. Too late now; if he misjudged the situation it would make things worse. Best to play by the rules.

Boris unlaced first one side curtain and then the other, his fingers freezing against the metal rings, burning on the rope. He followed her as she walked round, checking each pallet against the paperwork. This one was thorough. And that's when he noticed it.

The last pallet. The one Yuri had left behind. The one he'd gone back for. Tears in the stretch-wrap. Tape on the bags just under the tear. Why? Shit. To close the sample slits. Not just the top four bags, the decoy material, but all thirty-six bags underneath.

Suka.

His heart pounded at the realisation. The new engineer at Snow Science, the *kurva* had lied. She'd taken forty samples and only returned four.

For the first time in months he craved a cigarette.

The customs inspection finished, the *ošklivý knedlík* handed him his papers. 'Drive on.'

Boris pulled into the next lorry park and made the call.

Tuesday 1 March, Kranjskabel, Slovenia

Jaq lay on her bed and stared up at the magnolia ceiling. Low and featureless, a miserly skim of matt paint hid cheap woodchip. The hum of traffic swishing through slush penetrated the thin walls and echoed off empty spaces.

Her eyes strayed to the Paula Rego ceramic hanging above her desk, the only decoration in the room. The square tile, deft coloured brushstrokes on a cream-glazed background, was a gift from Great-Aunt Letitia.

A blue woman leaps over yellow fire, lithe and strong with well-defined calves and forearms. The bodice of her sleeveless dress is tight and smooth against a generous bosom; her ponytail and skirt soar in the air as she travels forward from right to left. Slip-on shoes balance precariously on outstretched feet as the ankles bend, the toes stiffen, the bridges arch to keep the mules from slipping into the fire. She glances down, assessing the danger, calibrating the risk, a half-smile on her lips. This is not a woman fleeing danger; rather one embracing it. Her hands swivel at the wrists, long fingers curled over to meet the tip of a straight thumb forming a dancing bud, ready to click in time to the campfire castanets, ready to flower.

The leaping woman is in complete control of the moment.

Vamos a isso! Jaq rose from the bed and selected some music, plugging her phone into the speakers. No one can sulk to P-Funk. As she bopped around the room, allowing George Clinton and Parliament to work their magic on her mood, she couldn't help noticing just how bare the room looked. She'd meant to buy more things, local handicrafts, soften it a bit, but she was always too busy at work. Work, ha! Some dream job, this. Look where it had led. Outmanoeuvred. Outnumbered. Outfoxed.

Jaq had been furious when she stormed out of Laurent's office. Now the anger was tinged with disquiet. She needed to talk with someone she trusted, but when she called Johan from her office at Snow Science, it went straight to voicemail. She left a message. Two missed calls from Gregor, the last person she wanted to talk to.

You're on your own. You are better on your own.

A low battery alarm interfered with Bootsy Collins's bass. When the track finished, she took the phone to the kitchenette and plugged it in. She selected something light, *Bledi Konj*, anything to distract herself, but Agatha Christie's words danced on the pages and she couldn't settle.

The phone rang. Johan? It took only two strides to reach the kitchenette. It wasn't Johan's number so she let it go to voicemail. Karel's deep voice emerged from the speaker, apologising for something. A little prickle of pleasure displaced her anger and she rang back.

'Hi.'

'Oh, you are there.' Karel's delight did something unexpected to her stomach. 'Can we meet up?'

Jaq hesitated. She definitely needed distraction. Why not someone unconnected with Snow Science or the mystery she had stumbled into? A couple of beers with a good-looking man. What was the harm in that?

'Where?'

'I'm in the area,' he said. 'I think I might be on your street.'

'What? On Korošca Ulica?'

'What number are you?'

She paused. Did she want Karel to know where she lived? It was meant to be a one-night stand, a moment of weakness: delicious, delirious, glorious weakness. Not to be repeated; she absolutely did not need another man in her life right now. One thing to wake up in the bed of a stranger, quite another to invite him into her personal space. This flat, however small and bare, was her private domain. Solitary. Claustrophobic. Cold. Bare. Mean. Miserable. *Bolas*, who was she kidding?

She told him the building and flat number. The doorbell rang almost immediately.

From the window she could see him, under a street lamp, a large brown paper bag in one hand, a rucksack on his back. He beamed up at her through the falling snow with a smile that was hard to resist.

And why resist? What the hell was she waiting for? She buzzed to unlock the street entrance and opened the flat door as he bounded up the inner stairs.

After the formal greeting, he kissed her on the mouth. Hot lips. Asking questions she wasn't ready to answer. Jaq withdrew and stared into his eyes. The deep blue took her back to the school laboratory, the time she used the complete stock of copper sulphate to grow beautiful spiky crystals.

She pulled away. 'How did you know where I lived?'

He cocked his head, a sheepish expression. 'You told me, just now, on the phone.'

'Then how did you get here so fast?'

He laughed. 'Nowhere in Kranjskabel is more than a few minutes away.'

He had a point.

'So, are you free for dinner?' he asked.

Snow was falling steadily, the wind whirling it into helical ribbons, a helter-skelter of icy white.

He followed her eyes to the window. 'I'm suggesting dinner right here.'

'The choice is limited,' she said. She'd stocked up recently on tins of baked beans and pot noodles, but European men were notoriously fussy about food.

'I brought some stuff,' he said. 'Can I cook for you?' He waited for her assent before unpacking a roll of sharp knives, a wooden chopping board and a wok from the rucksack. As he opened the brown bag, the smell of raw vegetables reminded her how bad her diet had become.

'I like cooking,' he said, and smiled at her, 'almost as much as I enjoy eating with someone interesting.'

'Interesting?'

'Yes, interesting.'

He could have said beautiful, but she would have rejected that. Beauty was a perfectly oval face, symmetrical features, a petite, hourglass figure, slender hands and small feet. Jaq's tall, skinny body and strong features had been unappealing to callow youths when it first mattered. Although not to Mr Peres. In fact, those early encounters with her chemistry teacher had saved her from wasting time with boys, time she had spent studying so she could beat them all. As she grew older and curvier, she became more comfortable with her body, secure in her own skin.

He could have said intelligent. She had been top of the class in everything she ever tried, from school through university.

He could have said nice. No, not if he had been paying attention. She wasn't going to repeat previous mistakes. She had learned her lesson, changed for good. Or for bad, depending on your point of view. Bad was the new good.

On balance, she would settle for interesting. In fact, she liked it.

He unpacked the food: a bunch of spring onions – long green stems with fat white bulbs; a stick of lemongrass; a stub of root ginger. He sliced the wizened and wrinkled skin of the ginger, revealing a buttery and glossy interior, and smashed the lemongrass with the handle of a knife, releasing a delectable aroma. For such a powerful man, he was unusually graceful, aware of his body, balletic in the confined space.

'So, you work at Snow Science?' he asked.

'Yes.'

'What do you do there?'

'Top secret.' She smiled. 'I could tell you, but then I'd have to kill you.'

He grinned, and she relented.

'It's a mix. Some long-term research, some operational stuff. Snow generation to keep the slopes open. Safety. We monitor the snowpack, the depth and crystal shape. Before it builds up to dangerous levels, we set off minor explosions to avert the danger of avalanche. The main research centre is in the French Alps, but there are satellite Snow Science labs all over Europe.'

'What are you working on right now?'

'Testing out different types of explosives: dynamite, ammonium nitrate, TNT, and so on, and the best form of delivery – in the air above the snow, on the surface or buried underneath. Quite often the only access is from a helicopter, so most of my work is on fuse timing. And then there's the more fundamental research on climate change and artificial glaciers and—'

'Artificial glaciers?'

'I can show you one day,' Jaq said. 'If you like.'

'I'd like that very much.'

Karel heated the wok and it started sizzling.

'And you?' Jaq asked.

'Nothing so exciting.' He sighed. 'Just a man for hire. A ski instructor in the winter and a mountain guide in the summer.' He splashed some wine into the wok and it bubbled and hissed.

'And occasional chef?'

'My training,' he said and smiled. 'But I prefer working outdoors.'

As they sat down to eat, her mobile rang.

She checked the caller ID. Laurent Visquel. Blast. 'Sorry,' she said. 'Work.'

He nodded sympathetically.

She answered. 'This had better be important.'

'Break-in at the explosives store,' Laurent said.

Merda. More than important.

'Come immediately and see what is missing.'

'I'm on my way.' Jaq jumped to her feet.

Karel sat back in his chair. 'You're leaving now? Can't you eat first?' he said.

'I'm sorry.' A knot formed in her stomach, twisting at her intestines. 'It's an emergency.' Another one. How many dinners had she ruined because things always went wrong at night? 'I have to go.'

He sat motionless, food untouched as she changed clothes and organised her bag. She avoided returning his gaze until she had one hand on the door handle. 'Please stay,' she said. 'I'll be back as soon as I can.'

'I understand.' He met her eyes. 'It's okay.'

Tuesday 1 March, Kranjskabel, Slovenia

Jaq's heart pounded as she raced down the back stairs to the drying room. She slipped into a warm jacket, laced up her stiff, dry boots and headed out into the night.

Screaming sirens and flashes of blue light, reflected and amplified by the crystalline snow, heralded the arrival of the police. As she stepped down from the shuttle bus she noticed Laurent silhouetted in the halogen lights. She gritted her teeth. The small, dapper figure cast an unusually long shadow, one that was in constant motion as he rubbed his delicate gloveless hands and stomped his feet in thin-soled leather shoes. He dressed as if he was still in Paris. An ambulance sped past in the opposite direction. Someone injured? She swallowed hard.

Laurent started shouting as she arrived. One of the uniformed policemen brought her up to speed. A break-in. The security guard, Stefan, overpowered and knocked unconscious. He hadn't seen or heard his attacker and remembered nothing. He sustained a nasty head wound and was distressed and confused. They had taken him to hospital to check for concussion.

The police were waiting for her at the door to the warehouse. Two uniformed officers stood well back, nervous at the proximity to explosives. Everything was wide open – the outside door, the inner door, the blast-proof cell doors, the metal cages and the sample cupboard. *Mau Maria!* What had been taken?

Laurent was still shouting. She ignored him. The older of the two police officers asked the questions: what time had she locked up? She answered, slapping the snow crystals from her hat.

'No sign of forced entry,' the junior officer said. 'Who has keys?'

Laurent opened the smart leather shoulder bag he always

carried, extricated his bunch of keys and rattled them. 'Two full sets, given only to those licensed by the inspector general to handle explosives.' He pointed to Jaq, and she produced the second set of keys from the secure pocket of her bag.

'What about the security man, does he have keys?' he asked.

Jaq shook her head. 'Only to the outer gate,' she said. 'He offloads the truck and leaves the material in the quarantine area. If these inner doors are opened and the alarm is not reset within thirty seconds, then the intruder alert goes off.'

'Who has the key to reset the alarm?'

'I do,' Jaq said, holding out the clock-winder key. 'And Dr Visquel.' She nodded towards Laurent and he held up an identical key.

'Well, let's see what's missing, shall we?' The junior policeman gave Jaq latex gloves and instructions to touch as little as possible until they'd dusted for fingerprints.

Box on box, shelf after shelf, case upon case, Jaq counted and recounted. She stopped and inspected. Everything was exactly as she had left it.

'Were the thieves interrupted?' she asked.

The senior officer shrugged. 'You have CCTV.' He pointed at the cameras. 'It shouldn't be hard to see what happened.'

After giving prints and a statement, Jaq hurried home. The flat was empty, no sign of Karel apart from a spotless kitchen.

Assailed by conflicting emotions, Jaq sat on the edge of the bed. The rush of relief – her own space, her own rules, time to think – mingled with disappointment. She laid her head in her hands. Get a grip, girl. You are better alone. Now is not the time to wobble. Blood sugar low. Eat something.

Karel had left several dishes in the fridge, covered in plastic film. She couldn't face cold fish but ate some fruit salad, crunching absent-mindedly on the small pieces of peeled apple mixed with soft banana and berries.

Should she call him? He could have waited. She stood at the window and surveyed the snowy street. Too late.

Why had someone broken into the explosives store? How had they broken in? Overpowering poor Stefan was not enough. Without a full set of keys and knowledge of the sequence, the alarm should have gone off thirty seconds after entry. Why hadn't the elaborate security system worked?

As she crawled into bed, something rustled beneath the pillow. Smooth paper with a faintly spicy scent. She held it under the bedside light.

A man came around. Gregor Coutant. Asked you to call.

The note was signed by Karel.

PS He says he's your husband.

Wednesday 2 March, Budapest, Hungary

A stream of cars accelerated away from the motorway tollbooth, a spray of grey slush fanning out towards the parked lorry.

Boris walked away from the policeman and lit a cigarette against the bitter cold. Had he made a mistake? What would Yuri have done? Damn. He never thought he'd miss that imbecile.

The Hungarian police officer had waved his lorry down just outside Budapest and directed him to a lay-by.

'What's up, mate?' Boris asked.

'Special control,' the official said. 'Wait here.' A radio crackled, and the official spoke into it.

Boris opened the door. 'How long?'

'They're on their way.'

'Who?'

'Inspectors.'

Stay calm. 'Why?'

'You've set off some sort of alarm.'

Just his luck. Mario would blame him for this, too.

It was lunchtime before a black car drew into the lay-by, followed by a minibus. Soldiers emerged and surrounded the vehicle. With guns. What the fuck?

A woman, dressed in a leather jacket, exited the car. Boris lit a cigarette and appraised the long black curls and motorcycle boots. Not a woman to argue with. She advanced towards the lorry as a soldier approached him.

'Papers?' The soldier demanded his documents.

'What's going on?'

The soldier inspected the papers. 'Wait here.'

The woman circled the vehicle holding a black box. Click … click,

click . . . click. Some sort of metal detector? Or a Geiger counter? *Hovno!* When she'd finished, the soldier handed her the papers.

'Open up,' she ordered.

'It's hazardous stuff,' Boris protested.

'I can read.' She tapped the clipboard. 'And I haven't got all day.'

Boris unlocked the back doors and flung them wide.

The woman passed the black box over the first pallet. Click . . . click . . . click.

'What are you looking for?'

She ignored his question. 'You picked this up in Slovenia, right?'

'Right.'

'And it's going to Russia?'

'To Smolensk. For recycling.'

Inside the truck, the device chattered like a machine gun.

The woman shook the papers at Boris. 'We'll need more information.' She turned to the soldier. 'Get me the number for Zagrovyl,' she said. 'They have some explaining to do.'

Wednesday 2 March, Jesenice, Slovenia

The market was in full swing. The thud of hessian sacks, the clatter of crates on concrete and the welcoming cries of stallholders hawking their wares drowned the chatter of shoppers.

As the doors of the bus opened, Jaq breathed in the aroma of roasting chestnuts: acrid charred shell and sweet white flesh. She moved quickly past souvenir stalls brimming with wooden carvings, oil paintings, straw dolls dressed in traditional costume, wicker baskets and glazed pottery, and scanned the stalls groaning under the weight of Slovenia's natural bounty: honeycomb cubes in wooden frames, liquid honey, jams and pickles in amber pots with bright red checked gingham hats, cheese and preserved meat wrapped in straw. At the fruit and vegetable stall she bought a small basket of fruit for Stefan – oranges, apples and grapes in a wicker tray wrapped in cellophane – and followed the signs to the Jesenice General Hospital.

The police had drawn a complete blank on the Snow Science break-in. There were no unusual fingerprints. The CCTV disk was missing, and the backup system battery had run down and failed to register. The security guard was the only person who knew what had happened.

As she got closer to the hospital, Jaq stopped and tried to call Karel again. No answer. So that was the end of that. She sighed. So be it. She certainly wasn't going to return Gregor's call.

Gregor Coutant. Her husband. Third-biggest mistake of her life. What was he doing in Kranjskabel? How had he found her address? And what the hell did he want?

She could imagine Gregor's reaction to finding Karel, apparently perfectly at home, in her flat. Even though she and Gregor were long separated, even though he was the one divorcing her, the discovery of a younger man in her apartment would have brought out the worst of his competitive antagonism. Ugly.

Let them fight it out. Right now, she had more pressing matters to attend to.

Jaq stood aside to let the mountain rescue team rush past with a stretcher. A figure lay prone inside a portable canvas tent. Sedated or badly injured – there was no sign of movement. The stretcher entered A&E, and Jaq took the next entrance, towards the main reception.

The smell of antiseptic hit her as she entered the ward. Stefan Resnik lay in an iron-framed bed in a ward of eight men. Suddenly much older, paler, thinner – and fast asleep. On a locker beside his bed sat a family photo, his reading glasses and a book.

He stirred as she sat down beside the bed.

'Hi, Stefan,' Jaq said. 'I brought you some fruit.'

'Whisky would be more bloody useful,' he growled.

'I'll remember that for next time.' She laughed. 'It's Jaq. From Snow Science. Came to see how you're doing.'

His eyes flew open. 'Dr Silver.' He scrabbled at the sheet, drawing it tight around his neck. 'You need to leave.'

'I'm sorry.' Jaq drew back. 'I didn't mean to disturb you.'

He sat up in the bed, suddenly flushed, chest heaving. A hand shot out and clutched her arm, preventing her from obeying his next instruction. 'Go away!' he said. 'It's not safe.'

'Not safe?' she repeated.

'Not safe at Snow Science.' He squeezed her arm and released it, his voice rising in volume. 'Get out while you can.'

A nurse bustled over, a middle-aged woman in a blue uniform and white lace-up shoes with fair hair scraped back from an angular face into a neat bun. Her low voice commanded attention.

'Madam,' she said. 'You are disturbing my patient.'

'I'm sorry.' Jaq tilted her head in contrition. 'I just arrived. I won't stay long.'

Stefan moaned. 'Get away!'

The nurse laid a firm hand on the chair. 'The patient has asked you to leave.'

'That's not what he meant,' Jaq protested.

'Mr Resnik needs to rest.' The nurse pointed towards the exit.

Jaq bent over the bed and whispered in Stefan's ear. 'Tell me what happened, please.'

'Shall I call security?' The nurse didn't raise her voice but spoke with an authority that brooked no argument.

Stefan closed his eyes and turned away. 'Go,' he whimpered. 'Before it's too late.'

Thursday 3 March, Budapest, Hungary

The rain froze to sleet as Boris shivered under a flimsy shelter in the motorway lay-by outside Budapest. They wouldn't let him back into the cab, nor would they let him leave.

Bad enough to have the wagon surrounded by soldiers, but he was cold, hungry and tired as well. He should have stopped for food before Budapest. Goulash. Dumplings. His stomach groaned with hunger.

When were they going to release the lorry? What if they arrested him? Demanded to know what was hidden in there? He imagined their faces as he told the truth. You think explosives are bad? Try some of this stuff! And what if he told them where he was actually taking the cargo?

For a moment, he was tempted. Trouble is, he'd done too many jobs already. If he told the authorities what he knew, he'd go straight to jail. And then Mario would find him. And send someone to kill him. Best to say nothing.

Something was up. A car arrived. A woman emerged. Christ, they were everywhere these days. Wide-hipped, short white hair, turquoise salopettes. Straight off a ski slope. None too happy about it. The other one, curly hair, leather boots, was back. She didn't look too chuffed either, a bunch of men in suits in tow. He sidled up, close enough to hear the introductions.

'Camilla Hatton, Zagrovyl.'

'Carla Rachman, International Atomic Energy Agency.'

The two women strutted away, out of earshot, and faced up to one another. They weren't going to see eye to eye. No siree. Spoiling for a fight. Phones appeared. Phones were exchanged. Curly stomped away, waving her free arm around as she listened, black curls bouncing with agitation. She practically threw the phone back at the white-haired woman, who caught it with a deft sideways feint.

The suits from Austria left with their noisy equipment in their fancy cars, and the soldiers piled back into their van. Victory! Zagrovyl one, bureaucrats nil. Sometimes it felt good to be on the winning side.

The customs official marched over and handed Boris the transport papers, signed and stamped.

'Okay, mate, you're good to go.'

'What was all that about?' Boris asked.

'A misunderstanding. It's all fine.'

Boris shrugged and started up the engine. As he passed the white-haired woman, he raised his hand in a salute. She stared straight through him.

Bitch.

Camilla Hatton. The name was familiar. The Spider would remember. The Spider never forgot anyone. That's what made him so dangerous.

Click. 47.45952, 18.99284. Intensity 72X, 648C

Boris changed gear and headed east. He would stop for food and rest as soon as he was clear of Budapest. Tonight, he would cross the border with Ukraine, and then he was on the home run.

Chernobyl.

Click. 51.389853, 30.094047. Intensity . . . Error . . . Error . . . Signal lost

Friday 4 March, Kranjskabel, Slovenia

Jaq pulled on her new snowsuit and patted the fabric to check for pockets. None: they had ignored her request again. She converted her bag to a backpack and coiled the climbing kit around her waist. The blasting planned for today had been delayed; the explosives store was still out of bounds as a crime scene. The ski slopes beckoned instead. A chance to forget about Zagrovyl and Gregor. Ex-employers and ex-husbands were bad news. Time to gather some snow samples.

Jaq rode the button lift up and skied to the crest of the north ridge. The portable laboratory in her bag contained everything she needed.

She anchored the orange nylon rope around a rock pillar, looped it through the metal karabiner at her waist and tied a double overhand knot. It was a smooth belay down to the site of interest.

The procedure was similar to taking samples in the warehouse. Extend the telescopic tube with a sharp point on the end, lock and push deep into the snow. When it was retracted, a core of snow came with it, perfectly representative of the layers below the surface.

She assembled a portable microscope and assessed the crystal shape in each layer, then extracted a second sample to measure density, filling a graduated tube, closing it with a stopper and leaving it in the sun to melt.

Six-pointed star snowflakes were stable; each unique and beautiful crystal had jagged legs that hooked and linked to another in three dimensions. The danger struck when supercooled water froze suddenly into pellets, rounded shapes that could easily slide one over the other. Avalanches were frequent on this sort of steep terrain. And the start and end of the season were the worst times.

The snow sparkled white and blue, diamonds and sapphires twinkling from the surface, reflecting the azure sky – a colour halfway between blue and cyan – a shade only possible in the thin, clean air of the High Alps. Completely absorbed in her task, unaware of the shadow moving across the snow until the warmth went out of the air. A tall figure skied up to her and blocked out the sun. Karel.

'I tried to call you.' Jaq shielded her eyes as she looked up at him.

'What are you doing up here?' he said. His voice was exasperated, unfriendly. 'The north ridge is dangerous, out of bounds.'

'Not any more,' she said, and held out the microscope. 'Look.'

But Karel didn't want to hear about snow crystals, dendrites and needles and rime and graupel. It was obvious there was only one thing on his mind. Gregor Coutant.

Jaq packed up. 'I'm sorry about Tuesday night.'

'Was anyone hurt?'

At least he had the decency to think of others before his wounded pride. 'The security guard is in hospital. But he'll recover.' She poked the snow with a ski pole and drew a circle in hollow dots. 'Gregor Coutant is my ex-husband. We've been separated for a while.'

She untied the knot and yanked at the rope. It slithered down the slope like an orange snake.

'I don't care,' he said, but his voice softened a little.

A good man. He believed this was an avalanche zone and still skied over, putting himself at risk to warn her. Perhaps there was still hope. Perhaps she could salvage something from the mess.

'Can I show you something?' she asked.

Jaq yanked at one of the plastic tubes buried in the snow. 'Do you know what this is?'

'No.'

'The artificial glacier project.'

Karel said nothing but straightened his skis and slid forward.

She pointed up at the mountain. 'The ice on the south side is melting. We divert the meltwater to the north and it freezes again. That way we preserve the fresh water and release it slowly over the summer months.'

Turning her back, she began to ski away.

'Science lesson over,' she called over her shoulder. 'Race you back.'

There were many reasons for taking the job at Snow Science. First and foremost was the need to get away. From Gregor, from Zagrovyl, from the lawsuit. Second was the chance to design and run her own experiments, try different combinations far from anyone who could be hurt. But the third and most uncomplicated was the chance to ski again.

The snowboard was handy for low-level commuting, but for the high slopes, she preferred skis.

Jaq gave herself over to the sheer pleasure of doing something she excelled at. Launching herself off the cornice, the snowy overhang formed by the constant wind between rocky outcrops, she sailed clear of the glacier and landed on powder. The acceleration took her breath away. Nothing to beat the buzz, the thrill of speed, the adrenaline response to danger all around, the exhilaration of freedom.

As the westerly sun warmed one side of her face, a rush of cold air froze the other. The snow glittered, crisp and granular. She flew across a crust of ice, hearing the crunch as it collapsed behind her. Straight down the slope, a thin layer of water making the hardpack faster. Careful. Approaching the treeline. Zigzag round the trunks, jump over roots. Flobble-flump. Rafts of snow fell from branches, bringing with them the scent of pine pitch. Swish, swoosh, sliding into the valley, glancing up at glorious jagged mountains tinted rose gold. Avoid the mashed potato on the southern dip, snowplough-turn to skirt the novice skiers. Slew, stem and glide as the resort came into view.

Jaq didn't pay much attention to Karel, but slowed a little on the run-out to allow him to overtake her as they carved into the valley.

With his hat and goggles already removed, he waited for her.

'What now?' A deeper question in his voice.

What indeed. What did she want from him? Nothing. Just for him to ski away and forget her. And yet.

'I think I owe you dinner,' she said. 'We can heat the fish.'

'Baah . . . old fish is no good.' But she could hear the smile. 'I'll pick up something fresh and be there in twenty minutes.'

They had undressed one another long before they finished the pizza.

When Jaq woke, Karel was still asleep. She scrutinised his untroubled brow, his square, clean-shaven chin, the long ringlets of hair and curling fair eyelashes: young, beautiful, the face of an angel.

She got up and wrapped herself in a towel to make breakfast. He deserved something better than instant coffee. At the back of the cupboard she located the Bialetti Dama – a Christmas present from Johan and Emma.

A noise rose from the bed. Karel was murmuring in his sleep. He had a beautiful voice, low and slow. She pressed one hand to her ear to control the melting memory of last night. The moment she'd stopped and looked at him, really looked. He was filling a glass of water for her, standing here, where she was now, at the sink. Naked. From the bed she'd studied the shape of him outlined in the moonlight, then listened to his approach through the darkness, tingling all over, aching for his touch.

A shaft of sunlight warmed her face through the kitchenette window. The skiers were already queuing at the lifts. White snow contrasting with the bright colours of their ski outfits: red, orange, yellow, green, blue, purple, turquoise. Turquoise salopettes. Damn.

Why had she agreed to meet with Camilla Hatton? Did Camilla really work for Zagrovyl? Could the company have changed so much? She didn't fit the profile of a Zagrovyl director. Survival of the meanest, the ones who succeeded were misanthropic, aggressive, thrusting bullies. There was something credible about Camilla. Which made the subterfuge even more disappointing. Camilla had tricked Jaq into leaving the mysterious samples unattended. Camilla had told her to keep her distance. Camilla could not be trusted.

What to make of Stefan's warning? The security guard had been shaken. Already in poor health, now he was confused. Paranoid even.

Cold brown water trickled out as she unscrewed the jug of the Bialetti Dama from the octagonal base. Ugh. Had she forgotten to clean it out last time? The tightly packed coffee grounds had turned green and an impressive colony of white spores had made a home inside the filter section. After scraping the residue into a bin, Jaq filled the sink with hot water and detergent and let her hands float in the warmth, the water caressing her skin as she scrubbed the coffee maker clean. She dried it and poured coffee into the funnel, packing it down with the flat side of a spoon before inserting it into the base and screwing on the jug.

It was a thoughtful gift. It must have been Emma who bought and shipped the little coffee pot in time for Christmas, but Jaq knew the choice of gift was Johan's. He always knew the right thing for her. If only she'd heeded him. Johan had warned her about Gregor. She'd chosen not to listen. Too late now. Johan had Emma and the children, and they were happy. A sigh from the bed reminded her that, today at least, she was happy too.

She lit two gas rings, placed the assembled coffee machine on top of one and a small pan of milk on the other. Three minutes. Behind her the sheets rustled and sunbeams danced as bare feet padded across the floor, bringing the scent of liquorice and the heat of a smooth chest pressing against her back.

The towel fell to the floor and she glanced down at her body. A hand cupped her breasts, her nipples erect even though his fingers hadn't reached them yet. The other hand wandered across her belly. She rubbed her buttocks against him, bringing her hands down to steady herself against the kitchen counter, pretending to continue with the coffee activities, placing the cups on tinkling saucers. His lips were on her neck, then his tongue in her ear whispering his request.

As he slipped inside her, three things happened at once.

The Bialetti Dama spluttered and whistled as the last of the pressurised water rose up through the ground coffee. The milk in the pan boiled over, a great wave of white froth cresting and foaming onto the cooker, extinguishing the gas flame so the rising steam carried the smell of fresh coffee, burnt milk protein and unburnt gas.

And outside, a massive explosion shook the whole valley.

Saturday 5 March, Kranjskabel, Slovenia

Jaq ran from her flat to Snow Science, racing towards the source of smoke that billowed and soared. Barely registering the cold; her calves and thighs burning with the effort of running uphill. As she neared the main gate, a police car roared past, soaking her in a salty grey spray. She weaved past two fire engines parked outside, all eyes on the pillar of grey smoke rising from the warehouse into the calm blue sky.

'*Spusti me noter!*' The fire chief raised his voice, demanding to be let in. Patrice, the security guard, stood firm, barring the entrance.

'Stop! *Ne vstopajte!*' Jaq yelled. Panting with the effort of covering the final few metres, she kept the information brief. 'Explosives. *Eksplozivi.*'

'I told them,' Patrice said.

'Well done.' Jaq turned as an ambulance arrived. 'Anyone hurt?'

'No one in today,' Patrice said and handed her the yellow hi-vis jacket with Emergency Controller, *Sili Glavni*, stamped across it in red.

Jaq bent forward with her hands on her knees and took a few deep breaths, before slipping the vest on and signalling to the fire chief to join her in the control centre.

She showed him the inventory and the sign-in sheet and asked Patrice to confirm no one had been working inside Snow Science at the time of the explosion. 'Can we take the fire tender to the rise and go up in your aerial platform?' Jaq pointed to the hydraulic ladder at the back of the fire engine.

From a vantage point twenty metres above the site, Jaq examined the roof of the warehouse. The cinder block walls were intact, but the plastic sheeting on the side and roof of one corner had been blasted away. The smoke was thinning. The worst was over.

'No sign of fire,' she shouted down.

The fire chief relayed the information to the brigade. Jaq closed her eyes and created a mental map of the warehouse. The stock of high explosives was on the south-west wall. All the damage appeared to be in the north-east corner, where there was only a small office, sample cabinet and a vending machine that didn't work.

'Bring me down,' Jaq said. When she arrived at ground level, she presented a summary. 'The event is confined to a small area of the warehouse. I'll wait until the smoke has cleared and then go in and check.'

'No.' The fire chief wagged a finger at her. 'Products of combustion can be every bit as dangerous as fire.'

'Oxides of carbon and nitrogen.' Jaq nodded. 'I know. I do this for a living. I'll test the atmosphere first.'

'Ignition temperatures?' he asked.

'Above two hundred degrees centigrade,' Jaq said. 'Biggest risk of detonation is percussive,' she clapped her hands to illustrate, 'and the shock wave has long passed.'

They sent a robot in first, a thermal camera and gas detector on wheels. Jaq moved back as they opened the warehouse door. A smouldering fire could be reignited by a rush of oxygen.

Patrice came up to her. 'Dr Visquel has arrived. He wants to see you in the office.'

'Once we're sure—'

'He said to come immediately.'

'Tell him you passed on the message.'

Patrice nodded and turned away.

As soon as the fire chief gave Jaq the all-clear, she made her way into the warehouse. As the smoke dissipated, the damage to the north-east corner became visible. Sunlight shining in where the corner walls and roof should be illuminating the tangled mess.

Bright red melted plastic mixed in with white electrical conduits, fragments of yellow foam insulation, shining steel and charred debris.

Only when she was sure that the dynamite and TNT safes remained undamaged did Jaq obey the summons from her boss. To tell him that it could have been worse. The ammonium nitrate was gone, but the segregation and blast walls had prevented any spread. There was no immediate danger; any risk of a secondary explosion had long passed.

'I'm taking over,' Laurent said.

Well done you. *Puxar a brasa à sua sardinha.* Let someone else do all the work, and then swan in to take the credit. Except that none of them could take any credit for this mess. They were both responsible for the safety of the explosives store. They had both failed.

Jaq handed over the incident controller vest. 'How can I help? What can I do?'

'Go home, Dr Silver.' Laurent's mouth twisted. 'I think you have done quite enough.'

Monday 7 March, Moscow, Russia

Frank stepped out of the car, leaving his assistant to pay the taxi driver. She scurried after him as he strode up the steps of the Moscow hotel, sliding on the ice despite her ugly, flat shoes.

'I don't need you for the next meeting, Shelly,' he said from the top of the steps. A uniformed doorman held open the heavy wooden door and Frank stepped into the grand entrance of what had once been a Romanov imperial palace: heavy on the marble and gold leaf, weak on plumbing and Wi-Fi speed. He preferred modern and functional, but Shelly had selected this ridiculous place.

'It's Raquel, sir.' She caught up with him, panting. 'My name is Raquel, Mr Good.'

'I don't need her either.' His upper lip curled with disdain. 'I will need the Smolensk Two project report in the morning, so make your own arrangements for dinner. I'll see you in time for the car tomorrow.'

Raquel had been an extraordinarily disappointing choice for this trip: prickly, aloof, bolshie, unwilling. Frank waited until the door closed behind her before acknowledging the man waiting in the shadows.

'Mr Good?'

The black-suited man shuffled forward with the stoop of someone who suffered with his spine; it made his thin arms and legs seem too long. Framed by spikes of black hair, decades of pain were etched onto his gaunt face. This must be Pauk Polzin. When Frank heard the accountant referred to as 'The Spider' at the Tyche meetings, he assumed the nickname came from the financial webs he wove, the traps he laid. Now Frank had to suppress a wry smile. The Russians were so literal; he'd never seen a man who more closely resembled a spider. 'You must be Pauk.'

The man nodded and held out a bony hand. His grip was firm but icy-cold. Frank led the way to a private room that opened off the main bar. The waiter filled shot glasses and then left them the bottle before closing the connecting doors.

'*Za vas!*' Pauk toasted Frank and then without further preamble, he asked, 'You need my help?'

The Spider came highly recommended; his perspicacity was legendary, his discretion assured. Frank paid his dues; he had not forgotten that the Tyche deal had been made possible, in part, by this man. A small part only, Frank had done all the real work on the deal, but one favour deserved another, at least now it suited him.

Frank sank into a padded armchair and sipped at his vodka, gesturing to Pauk to sit.

'I prefer to stand,' he said.

'But I insist.' Frank smiled, pulling unfamiliar muscles in his cheeks, stretching the skin. He observed the spidery man manoeuvre his long body, fascinated by the variety of grimaces that rippled over his sallow face as he folded, first at the knees and then at the hips, waiting until Pauk was perched on the edge of the velvet sofa opposite before continuing.

'How much do you know about the Zagrovyl factory at Smolensk?' Frank asked, refilling both glasses.

'All factories are the same to me,' Pauk said. 'Money goes out, money comes in.' He downed the vodka, reached into his jacket pocket and removed a slim packet of cigarettes. 'More money comes in than goes out, I am happy, I leave the factory alone.' He peeled off the cellophane wrapper and opened the lid, tapping it on the onyx tabletop to dislodge a black and gold Sobranie, which he offered to his interlocutor. Frank shook his head and waved a hand under his nose, pursing his lips with disgust. Pauk stroked the protruding cigarette with his fingertip before pushing it back into the packet and returning it to his pocket. 'More money goes out than comes in, I am angry, I fix it.'

'Good,' Frank said. 'They told me you were the right man for the job.'

'The job?' Behind thick glasses, Pauk's black eyes glinted.

'The Smolensk expansion project is late,' Frank said. 'Very late.'

'*Likha beda nachalo*,' murmured Pauk. 'Beginning is the big trouble.'

Frank stood and stretched. 'No.' He shook his head and waggled a finger to make his point. 'More than start-up problems.' He paced to the window. 'They made other promises that they haven't honoured.' He spun round to admire his tall, straight shadow projected by the slanting light onto the marble floor. 'They buy equipment but don't install it,' he said. 'Then they buy spares for the equipment they never installed.' He returned to his companion, standing behind him so Pauk had to twist to maintain eye contact. 'They pay for raw materials that get lost, make double orders, accept returns for recycling but never recycle it.' Frank clenched his jaw and his voice dropped to a whisper. 'Something doesn't smell right.'

'*Doveryai no proveryai*,' Pauk said. 'Trust but verify.'

Frank snorted. 'I prefer verification to trust.' He returned to the table and opened his briefcase, handing a red folder to The Spider.

Nicotine-stained fingers flicked through the pages. Pauk nodded. 'Yes, I will take this job.'

'Good. We are going to the Smolensk factory tomorrow. Can you join us?'

'Why would I go there?' Pauk asked. 'Everything is already in the numbers.' He tapped the red file with a long, yellowed finger.

Frank opened his mouth to protest, but Pauk held up a hand.

'Some people look at numbers and they see only numbers,' he said. 'I swim with the numbers, I dive through them. I see everything: what comes in and what goes out, the factory, the assets, the efficiency and inefficiency. The way to understand the people is through the numbers. I see the weak, the stupid, the lazy, the greedy, the brilliant . . .' he glanced up and smiled, '. . . and I always find their secrets.'

Frank relaxed in the armchair and crossed his legs. It was good to be out of England. Refreshing to be in a country where people called a spade a spade and knew how to wield the blade effectively.

Frank proposed another toast. As Pauk slammed down his glass, there was a knock at the door. Pauk shouted something, telling whoever it was to wait outside.

'I have seen your numbers, Mr Good,' he said. 'I know a little about what you like.' He opened his briefcase, took out an envelope and handed it to Frank.

Frank tore open the seal and removed two tickets for the evening performance at the Bolshoi: *Orpheus*, Stravinsky. The Russian Bach. A slow smile formed.

'You have excellent taste in music.' Frank studied The Spider with new respect. 'You will accompany me?'

Pauk shook his head. 'I have work to do.' He stood. 'But by chance a cousin is visiting Moscow from the countryside. She is a simple girl, but she loves the ballet.' He clapped his hands and shouted something. The door opened and a young woman in a fur coat appeared in the doorway.

'I am Nadya. I am for Frank,' she said. Frank licked his lips. A stunning cousin, no country bumpkin. Sultry red lips, dark, smouldering eyes under heavy kohl.

'Wait outside,' Pauk ordered.

Nadya pouted and winked at Frank before turning on her six-inch stiletto heels, wiggling her arse as she left the room.

'Perhaps you would do me the great favour of accompanying Nadya to the ballet?'

The evening was looking up, definitely looking up. The fur coat would be more interesting company than that frigid Raquel, or whatever she was called. What a criminal waste of a plane ticket; he wouldn't make that mistake again.

Pauk was rising from the sofa. Frank observed the man's laborious progress, fascinated by the symphony of pain communicated through a single gasp.

Once upright, Pauk collected himself quickly. 'I have ordered a chauffeur-driven limo for you,' he said. 'Perhaps a tour of the city bridges would help you become better acquainted with Nadya before the . . .' he coughed, 'main performance.'

Frank rose to his feet.

'Well, Pauk,' he said, 'you certainly do your homework.'

'Numbers.' Pauk tapped the folder. 'It is all in the numbers.'

Frank picked up his briefcase. Good. The meeting had gone well. The wheels were in motion. Time to relax and let others take care of the details. He nodded to the door. 'We can't keep a young lady waiting, now, can we?'

The fur coat sashayed towards them as they entered the lobby. When Pauk introduced her formally, she stretched forward to kiss Frank on either cheek and her coat fell open.

Frank wasn't sure what the dress code was for the Bolshoi these days, but he was sure it involved more than lingerie. Perhaps it was just as well Pauk had booked a private box.

Tuesday 8 March, Kranjskabel, Slovenia

The wooden floorboards creaked as Jaq paced up and down, waiting to be called to the accident inquiry. Was there anything worse than waiting? Part of the skirting was coming away from the wall and she deliberately clipped it with her heel at each turn. How many lengths of this windowless anteroom – a space as gloomy as her mood – before it broke free? She didn't complete the experiment; the moulded pine panel was still hanging on by a splinter when a door opened and Laurent emerged, his face twisted into a rictus of peevish irritation.

He barely glanced at her, waving his hand to indicate she should replace him in the meeting room. Her eyes followed where his finger pointed to the wintry glare of the accident inquiry chairman, seated at the head of the long oak table. From Paris, he had the right sort of face for an accident investigator – a stern, impassive mask and hooded eyes framed by a mane of white hair. Next to him sat Sheila, with a notebook. Flanking them were half a dozen members of the Snow Science European team, all based in France, most of whom she knew only by reputation.

Sheila smiled and indicated a seat, before passing a manila file to the chairman. He opened it and took out Jaq's CV.

'*Elle parle français?* He addressed the question to his neighbour.

'Yes, she does,' Jaq said in English, then switched to faultless French as she answered a series of questions about her professional qualifications, previous employment and activities at Snow Science.

The technical director cleared his throat. 'Tell us, madame, in your own words, your activities at the explosives depot on Tuesday 1 March.'

The day of the break-in. 'I came in early to prepare the explosives for blasting the north face.' Jaq filled her glass and sipped some water. 'I locked up and went to the laboratory to analyse some samples, but I was called away to meet a representative of our supplier, Zagrovyl.'

'We don't need to know operational detail,' the chairman said. 'Just focus on the security of the explosives store.'

'I think this might be important.'

The technical director whispered something to the chairman.

'We'll decide what's important,' the chairman said.

Jaq leant forward. 'Did you know that the samples I am referring to were removed and destroyed before I could complete the analysis?'

Sheila looked up and frowned at Jaq. A warning?

'Wait a minute.' The technical director stood up. 'Let me get this straight. Did you analyse samples from the two tonnes of ammonium nitrate that were in the warehouse at the time of the explosion?'

'Yes.' Rita ran the samples.

'Did you find any problem?'

'No.' Jaq had checked the analysis and signed them off.

'So the material was fully approved?'

'Yes.'

'Could a problem with composition of the material in the warehouse have contributed in any way to the explosion on Saturday 5 March?'

'In my opinion, no.'

'So, the samples that you refer to relate to some reject material that was delivered by accident on Saturday 26 February and removed two days later on Monday 28 February.'

'Yes.'

The chairman and technical director exchanged glances.

'Laurent told us about this incident, you can move on.'

'But—'

The chairman leant forward. 'I said, move on.'

Before Jaq could protest, the technical director resumed his questioning. 'Who locked up on Tuesday, before the break-in?'

'I did.'

'Where were you when the break-in happened?'

'At home,' she said. 'Dr Visquel called at about 8 p.m.'

'What did you do?'

'I went straight to the depot and checked the inventory with him.'

'Was anything missing?'

Jaq rubbed her chin. 'Everything appeared to be there.'

'Appeared?'

'The same crates, boxes and packages with the same serial numbers.'

The technical director stroked his white goatee. 'Why would someone go to all the trouble of attacking a security guard and breaking into an explosives store and then not take anything?'

Jaq did her best Gallic shrug. 'Perhaps they were looking for my samples.'

The chairman puffed in irritation. 'Dr Silver, your boss told us about the reject samples. He has also explained to us that they are of no consequence to this inquiry.'

Jaq started to object, but he raised a hand and spoke over her.

'The samples, and the material the samples represented, had already been removed. Laurent warned us you might bring this up. If you persist with this line of argument, I can only assume it is a tactic to divert our attention away from your responsibility for explosives safety and security.'

Jaq bit her lip. Laurent had already poisoned their minds against her.

The technical director resumed. 'What happened after the break-in?'

Jaq laid her hands on the table. 'The explosives store was cordoned off by the police while they carried out a forensic investigation. It was only released on Friday, the day before the explosion.'

'And did you check the security arrangements with the police on Friday?'

Jaq shook her head. 'No.'

The technical director pursed his lips and wrinkled his nose. 'Someone had gained access on Tuesday. Would it not have been wise to change the locks or put in some additional security measures?'

'Yes, it would have been wise, but I was not party to the discussion between Dr Visquel and the police.' *Take that, Laurent. Two can play at your game.*

The chair reasserted his authority.

'Between the time the police closed their investigation on Friday and the explosion at 8 a.m. on Saturday morning, did you go to the explosives store?'

'No.'

'Describe the security arrangements to the panel.'

Jaq explained how the security guard only had the key to the outer gate, which led to a small open-air courtyard. He didn't have the key to the inner doors or cages or the key to silence the alarm. She confirmed that there were only two sets of inner keys. She had one set, Laurent the other.

'Where are the keys kept?'

Jaq picked up her glass and emptied it, maintaining eye contact. Where was he going with this?

'Officially they're meant to be left in the safe in the office,' she said.

'Officially?' The chairman was sharp, finely attuned to the slightest qualification. 'And unofficially?'

Jaq held up her Tardis bag, in briefcase form today. 'We keep them with us at all times.'

It was easy for people who worked office hours to write procedures; they were not the ones who had to be at the depot before dawn to book out explosives to the avalanche prevention teams. Whatever the standard operating procedure said, it was simply not practical to keep the keys in a locked safe which was only accessible during office hours.

Sheila glanced at her phone, approached the chairman and whispered in his ear. As she sat down her eyes met Jaq's and flashed a message of . . . what? Surprise?

'I think we have heard enough for today.' The chairman addressed the assembled company in booming tones. 'The inquiry will resume tomorrow at eight thirty.' He turned to Jaq. 'Madame, please be here at nine o'clock sharp. In the meantime, you have a visitor in the small meeting room.'

Gregor Coutant was standing by the long window, his back to her. Medium height, boxer's build, wavy brown hair with silver threads – expensively cut to appear debonair. She paused for a moment, observing the familiar solid silhouette of her ex-husband as he spoke into his mobile phone. He might as well glue it to his ear for all he was ever without it. She waited until he finished the call.

'Hello, Gregor.'

He turned with a sharp cry and strode forward, holding out both arms. She stepped reluctantly into the embrace. Four awkward Parisian-style kisses followed. He held her at arm's length and scrutinised her face.

'Jaq! You look gorgeous,' he cried. '*Ravissante.* The alpine air must be doing you the world of good.'

Flattery will get you exactly nowhere. 'Gregor, what do you want?'

'I heard you were in trouble. Can I help? I know the chair—'

'No.' No hiding the acid in her voice. 'Gregor, why are you really here?'

'To see you.'

'Bollocks.'

He had the decency to hang his head for a moment.

'Cecile gave birth.'

'Congratulations.' Christ, did she even know her stepdaughter was pregnant? Had Gregor told her? Probably. Was that why he kept phoning? They had never been close; Cecile was almost sixteen when Jaq married Gregor. His daughter lived with her mother and always resented her father's new wife. It was not surprising they lost touch completely when Jaq and Gregor separated. But Gregor didn't look like a proud grandfather. He looked awful: unshaven, wrinkled, great bags under his eyes as if he hadn't slept for a month.

'Is everything okay?'

'Not really.'

His eyes filled with tears. Tears? Gregor? Ironman. Had he transformed into a sentient human being since leaving her?

'There are some problems.'

'For Cecile?'

'No.' He sighed. 'For the baby.'

Jaq took his hand and led him to a chair. She sat beside him and let him talk while outside the snow fell.

'Intensive care.' He launched into a long and rambling story about the difficult birth, the emergency resuscitation, the removal of the baby to a specialist paediatric hospital. 'They are still running tests, but it doesn't look good.' He shook his head. 'It doesn't look good at all.'

'I'm sorry.' She squeezed his hand.

'I'm not good at this,' he groaned. 'They won't tell me anything, won't let me do anything. Cecile and her mother. Will you come to France? Come to the hospital with me?'

Her mouth fell open in astonishment. Was he serious? His ex-wife hated her, his daughter resented her and they were in the middle of a family crisis when the only thing that mattered was the new baby. How could she possibly help the situation? Cecile was a daughter by marriage. What did that make Jaq? Did a stepmother become a step-grandmother? In a marriage that was soon to be dissolved? Could you be an ex-step-grandmother? And how could she possibly help? Wait. Who was this really about? 'Did Cecile ask for me?'

He averted his eyes and studied the floor. 'No, but—'

'Where have you been, Gregor? Since you turned up at my flat a week ago?'

'Skiing.' He covered his mouth and coughed. 'I had to clear my head.'

Bastard. So much for his concern about his daughter. She bit her lip. No point in starting a fight. He wasn't worth the energy. 'Go back to Cecile, Gregor. She needs her father right now, not me.'

'But I need you, Jaq.'

Just as she thought. This was all about Gregor. When had it ever been otherwise? 'No, you don't.' *You had your chance. You blew it.* 'You need to sign the divorce papers and move on.'

His upper lip curled into a sneer. 'You certainly moved on pretty fast.'

Aaah, the discovery of Karel in her flat had riled him. Good. Maybe now he had some idea of what betrayal felt like. It wasn't Jaq who had broken the marriage vows first.

'Deal with it.' Soft eyes belied the harshness of the words. 'Send Cecile my love. Let me know if there's anything practical I can do.'

'I'm so tired.' Gregor put his hands over his eyes. 'So weary. I need to sleep.' He looked at her through his fingers. 'Can I stay over tonight, in your flat?'

The cheek of the man. 'I'll book you a hotel room.' *Because, right now, I have my own problems to address.*

'So, you won't help me?' He shook his head sadly, as if he had always known she would let him down. 'In my hour of need. Then this was another wasted journey.' He stumbled to the door and cast a reproachful last glance at Jaq. 'God knows why I ever married you.'

The sentiment was mutual. How could she have thought this older man was the mature and reliable rock she could tether her chaotic life to? Just another man-child who viewed the world through the lens of his own needs, indifferent to the feelings of everyone around him. She said nothing.

'I should have known. They warned me you were a cold, unfeeling, unnatural bitch.'

She let the jibe wash over her. Gregor was ill-equipped to deal with this family crisis, and he was a father, a grandfather, in pain. No point in rising to the insult.

'Goodbye, Gregor,' Jaq said. 'Send my best wishes to your family.'

After he left, she stared out at the snow. Poor Cecile. As if the agony of childbirth was not enough to deal with, to then have the baby taken away for its own good. Jaq bit her lip and clenched her fists. She knew exactly what it felt like – a thousand miles and two decades away – and there was nothing in the world that would ever lessen the pain. God's punishment for her sin, they said. Jaq lost her faith right then.

Cars were moving across the gravel outside. Jaq was dimly aware of the tail lights by the little rays of crimson that danced and sparkled in her eyes. Do not cry. Not here. Not now. She clenched her fists, jammed them against her eyes and did what she always did. Locked it down. Locked it in.

When she uncovered her eyes, it was to see the chairman leave the building with Gregor, his arm around her ex-husband's shoulder. They embraced, and without a backward glance, Gregor got into his car and swept away down the long drive.

Wednesday 9 March, Kranjskabel, Slovenia

After a restless night, Jaq rose early and walked from her flat to Snow Science, the cloudless sky changing from soft pink to baby blue as she climbed. By nine o'clock she was waiting in the panelled anteroom, but it was another hour before she was called back into the inquiry.

A new member of the panel sat next to the chairman. A slight man, thirties perhaps, clean-shaven, green eyes and wispy fair hair. He was not introduced.

'Dr Silver.' The chairman fixed her with a frown of compassion. 'I understand you received news of a family emergency, yesterday. Would you prefer to adjourn?'

'No.' Jaq met his eyes.

He raised his eyebrows. 'Then are you able to answer more questions?'

She held his gaze. 'Yes.'

The chairman made a moue of disgust, as if he would have preferred her to break down in hysterics.

'In that case,' he shuffled his papers, 'we would like to go back a little further. Can you tell us about the events of Friday 25 February?'

Jaq stared at the ceiling and thought back. Four days before the break-in, eight days before the explosion. 'That was the day I prepared the experiment for the north corrie. We're trying out high-energy explosives, remote subsurface blasting to keep explosives out of the helicopters and—'

'Just the warehouse activities, please,' the technical director interrupted.

What had changed? Was she imagining a new, undisguised

hostility? Even Sheila, someone she counted as a friend, would not meet her eyes. 'I locked up at about 6 p.m.' No time to wash, barely time to change into her party clothes.

'Was anyone with you in the warehouse?'

'The blasting crew. I gave them refresher training on the safe handling of nitroglycerine.'

The chairman turned a page in the logbook.

'So where were your keys, for example, last Friday night?'

'With me. There was a delivery due on Saturday so I took the keys with me.'

'And where did you spend the evening?'

Jaq sighed. She should have seen this coming. 'What difference does it make? The keys were with me at all times.'

'Answer the question.' The chairman's voice had lowered, threatening.

'I went out with the Snow Science team.'

'With the keys?'

'Yes.'

'Where?'

'Karaoke City.'

A murmur from the panel.

'It was a late Christmas party, organised by Dr Visquel,' Jaq said. Two months late. Cheapskate. They must already know this.

'And then?'

Jaq sat back in her seat and put both hands in her lap. 'How is this relevant?'

'We have already interviewed Laurent.' The chairman consulted his notes and frowned. 'You drank heavily and left the party at Karaoke City on Friday night with a man who was not your husband.'

Good Lord, was there no end to Laurent's pettiness? Hell hath no fury like a boss scorned. The keys were in her bag. Her bag stayed under the bed. Could Karel have deliberately lured her back to his flat, waited until she was asleep and then taken the keys from her bag and copied them? It was possible, but hardly plausible. They barely slept all night, too engrossed in one another's bodies. Surely she would have noticed if Karel had left the apartment? And how many key-cutting shops were open in the middle of the night in Kranjskabel?

'Yes, I had a few drinks and left the party with a man,' she mimicked the chairman's language and spiteful tone, 'who was not my husband.' She spread her hands. 'Why do you ask me if you already know?'

The chairman glared across the table towards her. His mask of control slipped, fury and disgust bubbled close to the surface.

'I am asking the questions here, not you, madame. And unless you answer openly and honestly, I am going to record that you deliberately obstructed this investigation.'

Jaq curled her fingers and let her ragged fingernails dig into her soft palms.

'What do you want to know?'

'This man, had you known him long?'

'No.'

There was a meaningful exchange of glances between the HR director and the chairman. The rest of the panel scribbled again on their notepads.

'How long?'

'I met him that evening.'

More furious scribbling.

'I see. What is his full name and address?'

'I don't know,' she said.

There was a collective gasp, and every member of the panel stared in frank astonishment.

'Pardon?'

'We didn't do much talking.' And how. She smiled.

There was a distributed murmur. The commercial director, a short fat man, appraised her with a leer.

'And in the morning, Saturday, you went straight to work?'

The technical director coughed. 'Records show that the truck arrived at 8 a.m.'

'No.' Jaq shook her head. 'That is wrong. Stefan called at about 4 p.m. The truck was leaving as I arrived.'

More collective murmuring. Shuffling of papers. 'The delivery was booked in at 8 a.m.,' the technical director insisted. 'Here is the consignment note.' He waved the paper at Sheila, who fetched it and laid it in front of Jaq.

Jaq stared at the document. Saturday. Eight a.m. Stefan must have got the time wrong. It was definitely late afternoon when he called. She turned it over. Curious. The automatic timestamp confirmed it: 08h03m17s. And even more curious, delivery of twenty pallets, not two. She checked the batch numbers. There were eighteen more in the same series. Why had Zagrovyl delivered twenty pallets instead of two? When had the extra eighteen been removed? She had a sudden memory of the dark squares in the snow.

'There must be other transport notes for that day?' she said.

The technical director shuffled a pile of documents. 'No more deliveries, one return.' He passed her a sheet of paper. She recognised the name of the haulier. SLYV. The lorry she had seen disappearing down the hill as she arrived at 4 p.m. According to the consignment note, they should have picked up the wrongly delivered eighteen pallets. Except they had got it wrong. Loaded one that should have remained and left a reject pallet behind.

The technical director drummed his fingers against the pile of papers. 'Who had access to your keys?'

Jaq snorted in irritation. 'Did you ask my boss that question?'

The chairman referred to his notes. 'Laurent Visquel's keys were handed over to the police after the break-in.'

He looked over at the green-eyed stranger who nodded and said, 'On the morning of the explosion, Dr Visquel's keys were locked away in the police safe in Jesenice.'

So, he was a policeman. Good; there was something behind the incident that was far too serious to be handled by Snow Science alone.

'The keys,' the chairman insisted. 'Who had access to your keys?'

'No one,' she said.

'Let us return to Saturday morning.' The commercial director was addressing her. 'The delivery lorry came at 8 a.m., and yet you didn't arrive until 4 p.m.?'

He had a point. Why hadn't Stefan called her? Why had he waited eight hours before contacting her? Waited until after a new lorry had come to pick up the excess material? She opened her mouth to tell them again about the unusual pallet, the samples she had taken, the fact that they had been destroyed, when the commercial director interrupted.

'You slept in, perhaps? Or were detained?' He was practically salivating at the thought.

'I would like to register a protest at this line of questioning. I simply cannot see how my personal life is relevant to the accident.'

The chairman closed his file with a snap.

'And I would like to register my disgust that someone entrusted with the care of high explosives could be so lax in her personal habits and thus endanger the security of the depot.' He shook his head, a lank strand of hair falling across his face. He wrinkled his nose. 'Drunken parties. Shacking up with strangers. You are a grandmother, for goodness' sake!' He smoothed the hair back into place. 'See where it has led. Madame, it is my opinion that you are unfit to hold an explosives licence.'

Jaq jumped to her feet. 'How dare you!'

The collective gasp in the room grew from a whisper to a roar.

The chairman gathered up his papers. 'Dr Silver, you are hereby suspended from all duties pending a disciplinary hearing.' He addressed Jaq. 'Sheila will deal with the formalities, and then the police,' he nodded at the green-eyed man, 'would like a few words.' He rubbed his palms together, washing his hands of her.

Chair legs screeched against the floor, and the inquiry committee stood to leave.

'I'm sorry, Jaq.' Sheila couldn't meet her eyes. 'I need your pass. And your keys.'

In a daze, Jaq opened her briefcase and unzipped the inner pocket. She detached her two personal keys from the ring and slid the other ten over on her pass. Sheila entered all the serial numbers onto a form and handed a copy to Jaq before scuttling out of the room after the chairman.

Jaq remained sitting, her anger visible by the way she held herself bolt upright, unnaturally still, every muscle tense.

'Good morning, Dr Silver.' The policeman held out a hand. 'Detective Inspector Wilem Y'Ispe of the Specialna Enota Policije.'

The Specials. Slovenia's FBI. Jaq ignored his hand. She began to cough.

The inspector dropped his hand and picked up her empty glass. He filled it and passed it to her.

Jaq sipped the cool water, letting it soothe her dry throat, and reappraised him as he talked. Young for a detective inspector. Intelligent eyes. His spoken English was fluent, barely a trace of an accent. Educated abroad.

'Am I under arrest?'

'No,' he said. 'But I could use your help.'

Thursday 10 March, Moscow, Russia

Frank sat in the executive lounge of Sheremetyevo airport waiting for the flight to be called. He brought a glass of Octomore to his lips, savouring the smoky taste on his tongue. Ah! The Scots. Whisky was the only good thing to come out of their freezing, rain-lashed, dismal country.

Raquel's report from the Smolensk factory lay unopened on the table. Whenever he had something tedious to do, Frank ran through one of the Brandenburg Concertos in his head. He selected No. 6 today, a particular favourite thanks to the absence of violins. Frank started playing at an early age, becoming the finest violinist in the school. But his talents went unrecognised. The day Bradley was appointed leader of the school orchestra, Frank smashed his own violin. If he couldn't be top, he didn't want to play. He focused on the piano, gravitated to the harpsichord and finally the organ, where he could be leader, conductor and master of his own orchestra.

He tapped out a lively rhythmic opening, gliding into the first melody. Once he'd laid down the warp, the development began, weaving those golden threads of the weft, a sound tapestry on the loom. Now for the puzzle of the canon, the challenge, the resolution. He sat back with a sigh: glorious!

Frank stretched his legs. A large flat screen television above the bar flickered with the ticker tape of financial news, numbers running across the bottom of the screen. The Spider was right. Better than any drama, the story was in the numbers.

He drummed his fingers on the cover of the report.

An armoured limo had taken Frank and Raquel from Moscow to Smolensk. He wasn't flying fucking Aeroflot again – pilots high on aviation fuel, stinking peasant passengers and ramshackle planes. He'd planned to while away the journey getting intimately acquainted with Raquel, but the ice queen continued to resist his blandishments. He might have tried more forceful persuasion if the driver hadn't been a woman, and bigger than him. She was pug-ugly, with the sort of neck muscles that suggested a shot-put or javelin career prior to security. Women together were unpredictable, feral cats, scratching each other's eyes out one minute and ganging up on you the next. You just never knew when the hairy armpit sisterhood would club together and attack. He took a power nap instead, opening his eyes as wooded hills and snowy fields gave way to a walled city with gold and blue cupolas. The factory was visible several miles beyond, a ramshackle sprawl of sheds and towers wreathed in a pall of yellow smog.

Ivan came out to meet them. He'd made some effort to look like a company executive, suit and tie under his fur coat and hat, but he couldn't hide his background. An ex-boxer gone to seed, lines of anxiety spreading over his crumpled, misshapen features. A face assembled from spare parts, leftovers that didn't quite fit together. How many times had that nose been broken, that cheekbone fractured, that jaw dislocated?

In the conference room a photo of Frank breaking the ground for the production expansion took pride of place. It was a good picture – he could give Putin a run for his money when it came to Action Man poses. Frank had the advantage of youth and strength, which must have annoyed the Russian president standing next to him, grimly applauding. Ivan's misshapen face was obscured by the spade handle in the picture, but the familiar deep voice of the former boxer rattled on as he confessed that they hadn't started up the new production line.

Frank's voice was ice-cold. 'When were you going to tell me about the delay?'

'It's not my fault.' Ivan sounded as if he gargled gravel. 'The Roseboro control system didn't arrive. The supplier messed up the order and it went to someone else. We had to specify a new one, start from scratch.'

'Unacceptable.'

'We're working round the clock.' Ivan spoke in a slow voice, ill-suited to conveying a real sense of urgency. He'd been savagely effective before, precisely why he'd been selected to lead the project. This time he'd been promoted beyond his competence: the Peter Principle. He droned on. 'We have a limited number of workstations . . .'

'Get more,' Frank said. Was he the only person who could see the big picture? He grew so weary of engineers who lost themselves in the detail, couldn't see the wood for the veins of the leaves on the trees.

'Then we'd need more people to program them,' Ivan said.

Did he need to be spoon-fed? 'So, hire them,' Frank said.

Ivan let out a long sigh. 'They are specialists, hard to find, and anyway, we can only make changes one at a time.'

The question shot out like a bullet. 'Why?'

'Safety tests.'

'I don't care about fucking safety tests. Just make some fucking product.' Frank snapped his fingers. 'Doesn't matter how you do it. Do whatever it takes.'

'But Zagrovyl company policy—'

Frank ground his teeth. The company was infested with gold-plated, time-wasting standards written by idiots, head office pen-pushers who needed a detailed risk assessment to paint white lines in a car park. His jaw locked, and the words slipped through clenched teeth. 'You heard me. Start now. Test later.'

'It's too late to make any product this quarter,' Ivan said.

Fury rose from the pit of Frank's stomach and flared outwards. The Russians had lied. Ivan had looked him in the eye a year ago, shaken his hand on a promise to start up on time. Ivan had failed. Frank pressed a fist against the vein that throbbed in his neck, clenching until the knuckles turned white.

Ivan the terrible. Ivan the useless. Ivan would have to go, and his whole team as well. What was the proverb Pauk had used? *Beda nikogda ne prihodyit odna.* Trouble never comes alone.

While Frank sacked the Smolensk project team, Raquel collected the evidence for retrospective justification.

Now they were heading back to England, going back to ask some questions, to instil some discipline, to kick some ass, to ensure that no Zagrovyl project was ever late again. Fear, that's what really motivated people.

Frank freshened his drink and spooned a little caviar onto a blini, already loaded with cream cheese and smoked salmon. He popped it in his mouth, licked his fingers and opened the report.

His pulse quickened as he scanned the full list of missing deliveries to the Smolensk factory, more than he had realised, millions of pounds' worth of equipment and materials ordered but never delivered: a control system, a glass reactor and columns, stainless steel tanks. Several disputes were ongoing with suppliers who claimed the material had been dispatched. It never arrived, and yet the books balanced. Physical things were missing but the accounts showed nothing amiss. There was more to this than met the eye.

As he flicked back through the pages of high-value transactions, two company names caught his eye. Both beginning with S.

Snow Science – the name rang a bell. An exceptionally difficult customer, the complaints, returns and credit notes were much higher than their consumption, ten times, in fact. What did an alpine research centre want with so much stuff? And why so many fuck-ups?

And SLYV – why was a Russian transport company paying Zagrovyl and not the other way around? Since when did a haulier pay for the privilege of moving material across Eastern Europe? He flicked back and forth. He could find no record of any reject material from Teesside ever being received in Smolensk.

Frank looked up at the white ceiling and pictured the list from the Teesside warehouse. Yes, Snow Science was on that list too.

Deliveries from England, rejected in Slovenia, shipped to Russia but never received. Where had it gone? Where had everything else gone?

Should he show this new information to The Spider to speed up the investigation? Frank rose and paced to the window. A ground marshal waved coloured sticks to guide a plane onto the stand. No, let Pauk work for his outrageous fee; let him reach his own conclusions, find out what else lay beneath this thin carapace of respectable accounting. Raquel had uncovered something murky, something rotten.

The little prickle of excitement swelled and fizzed. Information is power. There was nothing he relished more than catching someone with their fingers in the till, and then destroying them. The hunt was on. Frank would do some investigating himself.

No tannoy announcements in the executive lounge: a pretty hostess sought him out and whispered that his flight was ready for boarding. Passengers here were treated as people, not cattle. As Frank rose, he noticed Raquel still obstinately sitting beside the reception desk. He could have used his platinum loyalty card to sign her in as a guest, but as he was travelling business class and she was only economy, he didn't see why he should. If everyone came in here, then it would hardly be executive any more.

He'd choose more carefully next time. He expected more of his female travelling companions than mere competence. When he got back he would definitely organise some bonding activities, some team building to break down barriers. Maybe a bit of mud and grime was what she needed to loosen up. Pain and fear might encourage a friendlier response in future.

Frank smiled to himself.

Thursday 10 March, Jesenice, Slovenia

Bright sunshine lit the valley as the bus from Kranjskabel emerged from the shadow of the mountains and turned south towards Ljubljana. Jaq tugged at a window, desperate for fresh air. The noise of water, gurgling down the steep slopes, cascading towards Lake Bled, heralded the start of the big melt. Spring was poised, fluttering in the wings, ready to chase winter away.

So that was that. The end of her association with Snow Science. One season in the Alps. Her brief career in avalanche control at an end. Her professional reputation in tatters. Again.

What next? She'd agreed to talk to Detective Wilem Y'Ispe in Ljubljana once the forensic results came back. The date wasn't fixed, but she refused to hang around waiting. The mountains loomed above the town, closing in on her, turning the once-beautiful valley into a suffocating trap. Movement. Action. Get out of Kranjskabel. Out of Slovenia. Go somewhere to think. To act.

She could fly to Lisbon, visit her mother in the convent nursing home. Try to convince herself that it was madness rather than hatred shining from those unblinking eyes. Then escape to the beach. *Praia fora de época balnear.* Deserted. Plunge into the ocean and bodysurf the Atlantic breakers. Sit by the water with sand between her toes.

Was Gregor right, for once? What had he called her? A cold, unfeeling bitch. Had she brought this on herself? Disobeying direct instructions from her boss. Ignoring advice from Camilla. Ignoring the warning from Stefan. Poking into things that didn't concern her. Concentrating all her energy to fight the consequences for her professional reputation while her extended family were going through turmoil.

In response to a sudden quiver of shame, she typed a message to Cecile.

Thinking of you. Can I help?

She could hop to France from Ljubljana, visit her stepdaughter. There might be practical things requiring attention, stuff she could do. At least it would give her a focus while she was suspended. But how to avoid Gregor? She shuddered at the thought of having to spend time with him and his other ex-wife.

Jaq leant her forehead against the glass. The imprint of the mountains, the jagged outline of the Julian Alps, were transformed into a spectrograph. She cast her mind back to the day she tested the samples. The first tests – before she met Camilla – were inconclusive; she had planned to run additional scans, but Laurent destroyed the samples.

From what she remembered of the printout from the analytical lab, the spectrographic fingerprint, the first white crystalline solid had a more complex structure than ammonium nitrate. If not NH_4NO_3, then what was it?

Urea? $CO(NH_2)_2$. The simplest answer. Another nitrogen fertiliser made by Zagrovyl. Somehow urea had been packed in ammonium nitrate bags. Incompetent. But hardly criminal.

Urea nitrate? $(NH_2)_2COHNO_3$. An explosive, just like ammonium nitrate, a substance easily manufactured by low-tech terrorists the world over. Take urea, a commonly available fertiliser, add acid and stir. But Zagrovyl didn't manufacture urea nitrate, not as far as she knew.

Spectroscopy was a powerful tool in the right hands. Bombard a substance with energy and see how it wobbles. Like identifying someone from the shadow they cast as they dance. But you have to find the right music to turn them on, and the right illumination.

Analytical chemistry had never been her forte. Too fiddly. Damn her curiosity! She should have left the testing to Rita.

Her phone pinged. A reply from Cecile.

Nothing you can do. It would only upset Mum. Lily is out of danger.

She has many challenges ahead but beautiful and loved. Dad still coming to terms with it all and being a total dick. Ignore him. Thanks.

So, the baby had a name at last. Lily. And Jaq wasn't needed. The relief made her sink back into the seat, away from the window. Cecile had confirmed her gut reaction: she would not be welcome in France right now.

She ran through the inquiry again in her mind, groaning aloud as she remembered the disappointment in Sheila's eyes. Even her friend thought her capable of gross negligence, guilty as charged.

Was Sheila right? Was it Jaq's fault that someone had blown up part of the explosives store? Had she given them both motive and means?

Someone had taken her keys. But how? They were always with her at work, never out of her sight. When she was doing fieldwork, they accompanied her in a zipped compartment of her bag. She kept the ten work keys on the same key ring as her two personal ones. She couldn't get into her bedsit without them.

Someone had copied her keys. But when? While she was asleep? She slept alone, or with Karel. He was cooking dinner when the break-in took place. And making love to her when the explosion happened. There was no way he could be involved. Apart from when she was with Karel, was there any time she had let them out of her sight?

Jaq slapped her forehead and groaned aloud. Last week. Tuesday morning. The meeting with Camilla at Café Charlie. Camilla taking her bag. Hanging it up for her, covering it with her ski jacket. A man with a briefcase brushing past. She closed her eyes, trying to picture the man. Think. No memory of his face; she hadn't got a good look at him. Had he snatched her bag from under the jacket while she talked to Camilla? Thrown it back under the hook before she left? Pressed the side of each key into some sort of special putty? Too old school. Scanned them with a 3D scanner? Could you fit a 3D scanner in a briefcase? Was Briefcase Man working with Camilla? An accomplice? Was the meeting with Camilla engineered in order to copy her keys?

What was it Camilla had said? *Keep away. Stay safe. Don't get involved. It could be dangerous.* Camilla knew something. More than she was telling. *Merda.*

Jaq rummaged in her bag until she found Camilla's business card. She dialled. No answer. No option to leave a message. She was about to cut the call when she noticed. Something was different. What? The ringtone: not the long European *beeep-beeep*, but a vibrating English *dring-dring.* So, Camilla was no longer in Slovenia. She was back in England. Where? The address on the card. Teesside. The last place Jaq wanted to visit. Far too close to home.

The Snow Science inquiry had been laughably biased. Jaq didn't fit into their stereotypes. A grandmother leaving a nightclub with a sexy young man. They would punish her for that. Well, she wasn't going to let them. Jaq's mouth hardened. She had fought this sort of accusation before; she'd fight it again. Not with emotion. With facts. And win. Again.

Destination: Teesside.

Mission: find Camilla.

PART II: PAVANE ENGLAND

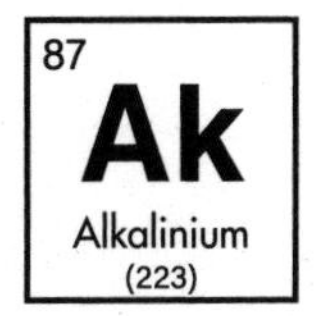

Friday 11 March, Teesside, England

The blue bridge straddling the River Tees loomed through the mist. Brown water oozed past a crumbling grey wharf underneath the dilapidated warehouse. Boris reversed the artic into the loading bay, swerving sharply as Mario emerged from the shadows. The swarthy Venezuelan yanked open the passenger door. Boris suppressed a cough as blue cigar smoke invaded his lungs.

'The Spider sent me.' Mario hopped into the cab with a face like thunder.

Boris shifted into neutral and tugged at the handbrake. The air brakes hissed.

'The Snow Science route is closed, *patrón*.' Unflinching, Boris met Mario's fierce gaze, impassive, glittering eyes almost lost under thick black eyebrows.

'What the fuck happened?' Mario growled.

Boris swallowed hard. 'Yuri took the wrong material. He left a precious consignment behind at Snow Science.'

'Yuri is a fucking imbecile.'

Boris made eye contact. 'Yuri *was* a fucking imbecile.'

Mario grinned, a thick-lipped leer that revealed teeth of impossible whiteness. 'You cleaned up?'

'Yes, *patrón*.'

Silver, the *kurva* from Snow Science, had lied to him. Thanks to the Hungarian customs inspection, he'd looked closely at the returned pallet. Sample slits in every fucking bag. Back at the Snow Science warehouse, he'd knocked Stefan around a bit to get what he needed. The samples were nowhere to be found, but he found something else, something more interesting. Hidden inside the vending machine.

Boris sat up straight, military-style. 'I retrieved the consignment.

Dealt with the witness. Destroyed the evidence.' He reached into the top pocket of his tartan shirt. 'But I found this.' He handed Mario a memory stick, a nub of hard plastic the size of his fingernail ending in a silver USB connector.

'What the fuck is this?' Mario held it up to the cab light. The black plastic glinted with a faint lilac sheen.

Boris extracted a sheaf of papers from the glovebox, computer printouts, tables with rows and rows of data, and handed it over.

240211 1845 54.597255, -1.201133, 800X, 0C,
250211 0608 51.126460, 1.327162, 152X, 648C,
250211 1145 50.966220, 1.862010, 152X, 648C,
250211 1904 48.585741, 7.758399, 152X, 648C,
2502112325 47.799400, 13.043900, 152X, 648C,
260211 0606 46.502800, 13.794400, 152X, 648C,
010311 0823 46.533200, 15.601100, 72X, 648C,
030311 1641 47.45952, 18.99284, 72X, 648C,
040311 0207 51.532153, 29.575247 Error, Error,
Error, Error, Error, Error, Error, Error,
Signal lost

Boris flinched as Mario whacked the pile of paper with the side of his hand. 'Explain.'

'I downloaded the information from the memory stick. Printed it out. It's data from a Tyche tracker.'

'How the hell do you know that?'

Boris detected a new note of respect hidden beneath the bluster. 'I used to work for Tyche, *patrón*.'

'Tie Chee? Who the fuck is he?'

'It's a company.' Boris spelt it out. 'T-Y-C-H-E. The bastards fired the boss, but not before he copied their ideas.'

'You worked with The Spider at this company . . . Tyche?'

'He hired me.' Both times.

Mario shrugged and leafed through the printouts. 'Where did you find these?'

'The memory stick was hidden in the Snow Science warehouse.' Boris locked his arms behind his head and stretched. Fortunately, the key was beside it; even so, it had still taken him a while to hack into the data. 'Someone has been watching us.'

Mario's olive complexion paled. 'Who?'

Boris shrugged. He had his suspicions, but he wasn't playing his hand yet.

'How long?'

Boris extracted the bottom page from the bundle and pointed to the left-hand column: year, month and day.

'*Mierda.*' Mario twiddled his moustache. 'From the beginning. How close did they get?'

Boris grabbed the top page. He entered the GPS coordinates from the third column into his phone and showed the screen to Mario. 'Signal lost at the border with Belarus.'

'*Bueno.*' Mario smacked his thick lips together. 'And you cleaned up at Snow Science? You're sure?'

'I'm sure.' Boris smirked. Amazing what a blob of Semtex could do in the right hands. 'But there is something else, *patrón.*'

Boris handed him the key he found with the papers. Silver, the shape of a bottle opener with a red spot.

Mario stroked the shaft with a nicotine-stained finger. 'What the fuck is this?'

'Hidden with the papers, but it doesn't open anything at Snow Science. I checked.'

A squall of rain rattled the roof of the cab. Mario turned on the cab light and carefully inspected the key. 'Any ideas?'

'A secure locker.' Boris puffed up his chest. Mario and The Spider were going to have to take him seriously after this. 'With the Tyche tracker hidden inside.' He pointed to the printout. 'Somewhere in the zone of alienation.'

Mario lit his cigar and stared out into the rain. He handed the key back. 'I need to think.'

Boris turned away to hide his smirk. Mario didn't know how to think. He was going to call the boss.

'You want me to clean up here as well?'

They relocated every few months, always one step ahead of the authorities. Fast and flexible. The hardest bit of the operation was getting the raw materials, but The Spider had friends with access. A little here, a little there, put the losses down to yield efficiency. Find a worker with a grudge. Bypassed for promotion. Moved sideways in the name of efficiency. Make him an offer he couldn't refuse. Exactly how Boris had started.

'Yes.' Mario opened the cab door. 'Clean up here, then stop all movements.' He swung his legs over the side. 'Nothing else crosses the border until you find the tracker.'

'Me?' Boris had been expecting something like this. Life was unfair. You gave management important information and instead of rewarding you, they gave you extra work. Dangerous work.

'Yes, you.' Mario jumped out of the cab. 'Find it. Destroy it. I'll make it worth your while.'

That was more like it. 'OK, *patrón*.'

Mario glanced over his shoulder. 'Do you need anything?'

Now, that was a first. 'I'll let you know.' Boris broke into a grin as Mario strode off towards his fancy car. Oh, yes. I need something. I need wheels like yours. A pay rise. More than that, I need you to count me in. Take me seriously. I'll find the Tyche tracker for you. But in return I want my share of the action.

Friday 11 March, Teesside, England

Home sweet home. Not so sweet. Jaq wrinkled her nose as the musty smell hit her nostrils. She dropped her bags and went to the window. Yarm High Street bustled with activity. The York train sped across the viaduct spanning the River Tees. A queue of cars formed behind a single-decker bus, the line of traffic at a standstill as an elderly gentleman climbed slowly aboard. Nothing much had changed in the time she'd been away. A town small enough to have escaped most of the chain stores and preserve its own local shops: a butcher, fishmonger, a couple of delicatessens, bars and restaurants.

Jaq slumped onto an old leather chair and closed her eyes. Slovenia felt like a bad dream. Any minute she would wake up and everything would be back to normal.

What was normal, exactly? Her life had never been what other people considered normal. She got off to a bad start and things went downhill from there.

She opened her eyes and let them wander over the Paula Rego ceramics hanging on the wall. Her Great-Aunt Letitia gave her one tile on each birthday and Christmas from the day Jaq turned eighteen. The last tile bestowed, not long before Letitia died, depicted a sculptress.

A woman, drawn in blue, sits and smokes, her legs wide apart, seat and thighs covered by full skirts. One hand is splayed over a knee, the other hand curls around the bowl of a long clay pipe. She peers through puffs of smoke, but is she gazing into the distance, or are her eyes unfocused: contemplating, reflective? Her hair is covered by a white headscarf and she wears a neckerchief neatly fastened at her throat with a square buckle that matches her hooped earrings. A solid, tidy, competent woman, an artist who has just created something pleasing: a woman proud of a life well lived.

Jaq jumped up and threw open the sash windows. Her fingers ran over the music shelf until she found what she needed. The polystyrene box snapped open and she inserted the disc – a polycarbonate base overlaid with a thin layer of aluminium topped with transparent acrylic varnish – into the stereo and pressed play.

As Ella Fitzgerald sang 'T'aint What You Do (It's The Way That You Do It)', a flutter of hope flew in on the spring breeze. Aunt Lettie's flat held Jaq's happiest memories. Peaceful. Safe. A waft of lavender from a tiny cushion brought with it her great-aunt's voice. *Come on, girl. No use hanging around. Dilly-dally, shilly-shally never buttered no parsnips. Seize the day. It won't come again.*

Camilla. She needed to find Camilla.

The hinges on the garage door squealed with rust, but Aunt Lettie's ancient Land Rover started first time. Jaq rattled along the cobbled back alleys, parallel to the high street, joining the main road at the bridge, and headed north. The Zagrovyl sign loomed over the flyover, a virulent blue casting its own shadow over the dual carriageway. Jaq slipped into autopilot as she entered the car park, turning left into the staff section, past hundreds of white rectangles separated into rows by neat box hedging until she arrived at number 179. Someone was in her space. She bit her lip and frowned. *You don't work here any more.* Jaq gripped the steering wheel, turned the car round and found the visitors' section.

She turned off the engine and remained in the car. A few changes – three charging points for electric vehicles, an empty chrome and glass bicycle shed, some hanging baskets and planters. She chuckled to herself. Zagrovyl's attempt at green credentials. All for show. Time to ask some questions of substance.

Her hand trembled against the door handle. This place held so many bad memories. She swallowed hard. Just open the door and stride into reception. Ask for Dr Hatton. How hard could it be?

And what if they recognised her? Fat chance. There was almost no one left from the old days. All the good ones long gone, sacked by the macho men. Shoot first and ask questions later. It was unlikely anyone here would remember her. She had rarely come to headquarters after the acquisition of her ICI factory by Zagrovyl. She clenched her fists. Time to demand some real answers.

The glass doors opened with a whoosh. The carpet was new, thick pile in the same blue as the company logo. A bad choice; it was already showing marks from muddy boots. The young woman behind the reception desk looked up; an unfamiliar face, a mask made rigid by too much make-up. What had happened to Pam, the smiling, kindly receptionist? Probably too motherly to fit with the dynamic new image.

'Good morning, may I help you?'

Jaq approached. 'I'm here to see Dr Hatton.'

'And you are . . .?'

'Jaqueline Silver.'

No flicker of recognition. 'Please take a seat.'

So far, so good. Jaq sank into the nearest chair and tried to regulate her heartbeat by distraction. She contemplated the changes. No expense had been spared in the new waiting area. Black leather sofas and chairs – strewn with bright blue scatter cushions, the Zagrovyl logo embroidered in gold – arranged around low glass and chrome tables. The hard lines of the modern furniture softened by an array of tall green ferns in pots, lit by a revolving display cabinet. She flicked through the journals on the table. A dismal selection.

The chair was deceptively uncomfortable, insufficiently padded, the metal frame pushing against her coccyx. Jaq stood and stretched before inspecting the cabinet. Shelf upon shelf of awards glittered under halogen lamps as they slowly rotated. Silver trophies – Queen's Award for Export; brass plaques on rosewood – Investors in People; engraved glass blocks – Six Sigma awards; Perspex sculptures – carbon footprint reduction. Jaq suppressed a snort of irritation. Easy to reduce your carbon footprint by closing all your manufacturing and exporting it to countries with lax environmental regulation. Net result – better for English lungs, worse for the planet. Where did the judges imagine the Perspex came from for the trophy? Their arses? She stroked a frond of the fern. Even the plants were plastic.

A leaflet boasting Zagrovyl's support for environmental projects caught her eye, the cover picture a sparkling, turquoise pyramid of ice: the Artificial Glacier Project. That explained a lot. Zagrovyl was a principal funder, along with EPSRC, Fustington Industries and NASA. Laurent wouldn't jeopardise the relationship with a major funding partner. He would do whatever Zagrovyl wanted him to do.

But what was it that Zagrovyl wanted?

A young man in a suit and tie pushed through the turnstile from the inner building, said a few words to the receptionist and strode over to Jaq.

'Good morning.' He extended a hand. 'Miss Silver?'

'Jaq.' She shook his hand. He could easily have been a model, with his high cheekbones, square jaw and sharp suit.

'You asked to speak to a Dr Hatton?'

'That's right.'

'I'm afraid it won't be possible.'

Jaq raised an eyebrow. 'Why not?'

'She doesn't work here,' he said.

Jaq stepped back and planted her feet apart. 'Then where does she work?'

'There is no one called Camilla Hatton working for Zagrovyl,' he said.

Jaq reached into her pocket and pulled out Camilla's business card. 'Then how do you explain this?'

As the young man examined the card, a flush rose from his neck and spread to his cheeks. 'Please excuse me for a moment.' He raced through the turnstile, phone already to his ear.

Jaq paced up and down, unable to settle. She filled a plastic cup from the water cooler. Nice clear polyethylene terephthalate cups. She drank two cupfuls, but her mouth remained dry. The receptionist answered the phone and then tapped away on a computer, casting sidelong glances at Jaq.

An aerial photograph covered one wall: Zagrovyl's empire in Teesside. The photo had been cropped. One site missing. Seal Sands. Just a few miles away to the north. A few miles, several light years and many unnecessary deaths.

The return of Mr Cheekbones jolted Jaq back to the present. Beside him stood a man in a blue three-piece suit, his waistcoat fastened with pearl buttons, matching cufflinks peeking beneath his jacket sleeve as he shook her hand.

'This is Frank Good.' Cheekbones introduced him in tones of reverence. 'European operations director for Zagrovyl.'

'Come.' Frank waved a hand, indicating a small glass-fronted meeting room beside the main waiting area. He closed the door behind her, leaving Cheekbones outside, sat at a hexagonal table and gestured for her to sit opposite. 'Perhaps you can tell me what all this is about?' He smiled.

The curl of his lips didn't match the expression in his cold, glittering eyes. It was as if someone had taught him which physical muscles to move to achieve the necessary rictus of mouth and cheek, but not what feelings should trigger it, a simulacrum of a real emotion. The smile of a snake.

'I was hoping you could tell me,' Jaq replied evenly. 'I need to talk to Camilla Hatton urgently.'

'I'm afraid you're mistaken. We don't have anyone by that name here,' he said.

'So, she has left the company?'

'No, she never worked here.'

What? Had she been hoodwinked by the tall, pale stranger in turquoise salopettes? Anyone could copy the Zagrovyl logo from their website and create a false business card. Why should she believe Camilla? Jaq blinked, as if it might shift the film over her eyes, the haze blurring and obscuring the truth.

Why not believe Frank Good, this clean-cut captain of industry with a natty taste in waistcoats? She made eye contact. Why? Because he was lying. They were both lying. The boy with the cheekbones claimed not to recognise the name, and yet he had known Dr Hatton was female. Most people associated the title 'Dr' with a man by default. And he used her first name, Camilla, before Jaq did.

Frank claimed Camilla Hatton had never worked here. Zagrovyl had acquired hundreds of companies with thousands of employees. And then sacked two thirds of them. How could Frank possibly know the names of all the ex-employees of all the Zagrovyl subsidiaries? He hadn't recognised Jaq's name, and she had worked here.

They were all lying. She was absolutely sure of it. So, what had Zagrovyl done to Camilla Hatton? And what might they do to Jaq Silver if she asked too many questions?

Time to go.

Jaq made her excuses and marched past the reception desk, through the sliding doors and back in the cool, damp air. Drops of rain fell onto her hair. She looked up and let the soft drizzle wash away the lies.

This had been an ICI office once: a division where engineers were listened to; an environment where dispassionate data trumped power-distorted emotion; a culture of respect and meritocracy; a commitment to new technology and lifelong learning. She'd been supported through her PhD, allowed to continue at Teesside University one day a week as an industrial lecturer, mentored to become an expert in her field of process safety, rewarded and promoted.

And then Zagrovyl took over. That's when it started: the changes, the headaches, more changes, then deaths.

Tony was the first man to die. In his mid-sixties, overweight and unfit, he had resisted all proposals for retirement, joking that he felt better at work than at home. He collapsed on shift one day. By the time the ambulance arrived, he was already dead. It was awful, but such things do happen out of the blue.

Nobody made the connection until Adrian collapsed a few days later. Co-workers administered CPR, he made it to hospital, but the damage was done. After many agonising months, the family capitulated and allowed the life support machines to be switched off. By which time there had been a third death, Peter, and the connection was undeniable.

That's the trouble with change. As Donald Rumsfeld put it, there are known knowns – things we know we know . . . known unknowns – things we do not know. There are also unknown unknowns – the ones we don't know we don't know.

But Donald missed out an important category: the unknown knowns, things once known but then forgotten.

She'd survived the suspension, the deposition, the inquiry. The HSE decided not to prosecute. Then Zagrovyl fired her. Oh, they masked it as redundancy, but there was no doubt why they wanted her gone. She was given no support when the families brought a civil action against her.

Raindrops streamed across her cheeks. Salty raindrops. It still hurt to come back here, to be forced to remember everything she would rather forget.

Jaq swallowed hard. How dare she wallow? How dare she complain? Didn't the bereaved families deserve some redress, some compensation, some closure? Even if they were fighting the wrong person?

She wiped her eyes with the backs of her hands. What's done is done. Never look back. Lock it down. Lock it in.

Radio TFM reported congestion on the A19 flyover. Jaq made a detour, driving to Port Clarence. She joined the queue of lorries waiting for the moving platform of the Transporter Bridge to cross the River Tees.

The platform docked and the barrier opened. Jaq drove into a bay and got out of the car. She stood on the platform and watched the river speed past.

Through the rain she admired the Cinderella of civil engineering, a slim-thighed blue maiden hidden away in a forgotten corner of Teesside among abandoned warehouses and empty docks.

Although not completely abandoned. Rubber screeched on tarmac as one lorry after another drove out of a ramshackle warehouse. Signs of life. Perhaps there was hope for Teesside. New investment. New beginnings.

But not for Jaq. There was nothing for her here. Nothing for her back in Slovenia. Not unless she could find Camilla. Prove that Camilla had tricked her. Copied her keys. Entered the warehouse looking for samples that no longer existed because Laurent had already destroyed them. Set off an explosion to make sure no evidence remained.

Had Jaq given too much away in her confrontation with Laurent? Had she been a fool to visit Zagrovyl here in Teesside? A gazelle bounding into the jaws of a lion. *Caramba.* Time to start being more careful. The bastards at Zagrovyl knew she was searching for Camilla. Would they try to stop her?

Friday 11 March, Teesside, England

Frank remained in the meeting room after Jaq left. His hands formed a steeple. One foot tapped the floor as he examined several courses of action. It didn't take him long to discard most of the options. More information was required for further refinement. He called HR, gave them two names. Then he called security and logged an incident report.

While he waited for the HR director, Frank meandered through a circle of fifths in his head, tapping out the Brandenburg Concerto No. 4, inventing a harpsichord part. Even Bach made mistakes sometimes. His fingers flew across the desk, stopping abruptly as Nicola knocked at the glass door.

'Well?' He didn't invite her to sit.

The HR director paused to catch her breath. She was not only repulsive, but woefully unfit. Compulsory corporate boot camp, that's what Zagrovyl needed. He would restrict access to the Zagrovyl canteen to those who could fit through a narrow door, and link the chocolate machine to a weighing scale – only those below a certain BMI would be permitted to withdraw the Snickers and KitKats they craved. That would soon get rid of this one.

'Dr Jaqueline Silver,' Nicola said. 'Former employee of ICI. When Zagrovyl took over the Seal Sands site, she was technical manager.'

'So, she works for us?'

'Worked. She left, after the Seal Sands incidents.'

'What incidents?'

Nicola looked as if she had swallowed a frog. Her eyes bulged and the wattle on her thick neck wobbled as she swallowed repeatedly. Her voice came out as a croak.

'The fatalities at the Seal Sands site.' Her protuberant eyes, eloquent with pain, spoke of shock that he didn't remember. 'When the authorities decided not to press charges, the families mounted a private prosecution.'

So, Nicola did have a soul after all. He'd found a way through those shithouse rat eyes to a gloopy, bleeding heart. A leader has to know what makes his people tick. Now he knew Nicola's Achilles heel, he could find the buttons to press. This was a discovery to save for later. 'Unfortunate, most unfortunate.' Before she could comment, he added, 'What about Camilla Hatton?'

Nicola shook her head. 'I've no record of any such person ever working for Zagrovyl.'

Good, Bill had done his work well.

'Is that all?' Nicola opened the door to leave.

'Wait.' He held up a hand. 'Do you know what Jaqueline Silver is doing now? Apart from making a bloody nuisance of herself.'

'I'm your HR director.' Her wrinkled nose suggested a bad smell in the room. 'Not a private investigator.'

Which gave him an idea.

Friday 11 March, Teesside, England

Boris did a double take. *Suka.* He crouched behind a stack of pallets and peered at the moving platform speeding away across the River Tees. Green Land Rover. Tall woman standing beside it. No doubt about it. Instantly recognisable. Silver. What the fuck was she doing in Teesside? What the fuck was she doing alive? Hadn't she triggered his trap? Set off the explosion? Been blown to bits? What had gone wrong? If it wasn't Jaq who detonated his little surprise, who the hell was it?

She was staring back at the warehouse. He ducked. Had she seen him? What did she know? Christ, was she on to them already? Too smart for her own good, that one.

Boris ground his teeth.

He'd promised to clean up. He'd promised it was all under control. If Mario found out, his life was over. SLYV did not tolerate failure. Mario had special punishments. Yuri's fate was a trip to Dignitas in comparison.

Káča pitomá! Time to take decisive action. Silver was a witness. Witnesses had to be taken care of.

Friday 11 March, Teesside, England

The gym sat above the Tees barrage, the weir that separated the river from the sea. Before its construction, melting snow from the Pennines clashed with high tides from the North Sea, leading to widespread flooding from Middlesbrough up through Stockton and back to Yarm. Now the drop from river to sea was carefully controlled, with a sluice gate that fed a white-water rafting course used by those seeking controlled thrills: an oxymoron if ever there was one.

Frank blasted his horn and accelerated towards a gaggle of idiots in wetsuits and fluorescent buoyancy aids dragging their kayaks across the road. He was early for his meeting on the squash court, but roads were for cars, not for aquatic arseholes.

Frank always used the same private investigator. His first divorce threatened to cost him a fortune until Bill came on the scene and fixed things. Intelligence was the key. Once you had the facts, the connections, then you could decide how to spin things to your advantage. Information was power.

And Bill was adaptable. He offered more than just surveillance. When called upon, he could make people vanish, people like Camilla Hatton.

Dr Jaqueline Silver wasn't going to be so easy.

As he was parking the car, his phone rang. Country code thirty-three. France.

'Bonjour, Monsieur Good. It's Monsieur Barré here.'

The new marine surveyor, the third expert to inspect *Good Ship Frankium.*

The yacht gave him two thirds pleasure and one third pain. Pleasure because he looked so good at the wheel. Pleasure because he had taken it from a rival down on his luck. Negotiating hard, he drove the price down to rock bottom, the former owner weeping openly by the time he signed over the registration document. A fine day's work.

Pain because of the ruinous expense of running it – not just the mooring fees and crew salaries – because underneath the beautiful interior writhed a viper's pit of trouble, fine cracks spreading through the hull.

'Well?'

'Repeated grounding . . . poor-quality patch jobs . . .'

Within two minutes Frank had stopped listening.

Did a marine surveyor ever call to give good news? It's all fine, Mr Good. The hull is going to get better all by itself. No need to go into dry dock, scrape barnacles, lather on eye-wateringly expensive paint from stem to stern. No need for complicated repairs, bigger bilge pumps, extra batteries or new sails.

Good news from Barré was about as probable as his Zagrovyl bonus paying out this year. The one year he really needed the money, the fucking useless Russians had fucked up.

'I estimate the cost of repair at . . .'

Frank's jaw dropped. Barré was just another money-grabbing bastard. Pah! He could go fuck himself.

One more season of fair-weather sailing, then he'd spruce it up and pass the *Good Ship Frankium* to some other unsuspecting bastard.

Frank cut the call, grabbed his kit and headed towards the gym.

Bill had cultivated the art of blending into the background. There was nothing memorable about him. Medium height, average build, clean-shaven, brown hair neither short nor long, regular teeth but not too white, straight nose but not too large. Lips neither thin nor thick. Hooded eyes. A face you forgot the moment he turned away. Even his sportswear was unbranded: white shorts, socks and trainers, a blue T-shirt and a grey racket. Totally nondescript.

It was easier to talk on the squash court. Frank hit a few balls to warm them up. 'I need some information,' he said, slamming the little black balls into the wall. 'On a woman.'

Anyone else would have made a smart-arse comment, but Bill was not one for small talk or innuendo. He returned the balls one by one. 'Name?'

'Jaqueline Silver.' Frank selected the ball he wanted for the match and put the others in his pocket.

Bill moved to the back of the court, bent his knees and rocked onto his toes, ready and waiting. 'What do you know?'

Uuuh. Frank served. 'Late twenties or early thirties. Sporty. Ski tan.'

Bill volleyed. *Thwack.* Frank hit a perfect boast off the front wall onto the nick. *Clatter.* Bill missed it, his racket hit the wall and skidded across the floor.

'English. Educated. Engineer.' Frank didn't wait for Bill to recover his position; he served underarm to win the next point.

'Ex-Zagrovyl,' Frank continued as he set up a rally. 'Asking awkward questions about a certain' – *wallop* – 'Camilla Hatton.'

'I see.' *Smash.* Bill sent a drop shot over Frank's head. 'I thought that particular threat had been' – Bill grunted, ran towards Frank's return and missed – 'neutralised.' He rested his hands on his knees and panted.

Frank served and rallied. 'It appears there are some residual issues.' He made a fist pump as Bill conceded the first game. Bill played a mean game of squash. Not quite as well as Frank, but a worthy opponent.

Bill served overarm. 'So, what do you want me to do about this Silver character?'

Blam. Frank returned the shot. 'Just information.' *Slam-slap.* 'For the moment.' *Swish.* 'Anything I can use against her.' *Squelch.* 'Skeletons in the closet.'

Bill sent a firm drive into the far corner and jumped aside as Frank rushed for it. 'How deep do you want me to go?'

Frank scooped the ball from the corner and volleyed. 'Something tells me you won't have to dig too far.'

Saturday 12 March, Teesside, England

Fine Georgian town houses lined Yarm High Street; the upper floors with their high ceilings and tall sash windows remained residential, but lingerie boutiques and cocktail bars had invaded the ground floors. Coming back from her morning run, Jaq glanced into the window of a new nail bar as she jogged past.

Camilla's nails. Perfectly manicured. Camilla's hair. Expertly styled. If Camilla had lived and worked in Teesside, she might have used a local salon. People talked to their hairdressers. Built up relationships. And there was one person who had direct access to every hairdresser in the North-East.

Natalie ran her salon from the side room of a barber's shop. Simple and functional: sink, chair, mirror, kettle and a large black-and-white photo of Paul Newman beside a window looking on to a side street above Yarm. Businesslike, down to earth, no frills. Just like Natalie.

Jaq didn't bother calling ahead. It was a short walk up the hill. She was greeted with a rib-snapping hug. 'Christ! Look at the state of your hair!' Natalie said.

'Can you fit me in?'

'You're in luck,' Natalie gestured to the empty chair. 'First client cancelled.'

Jaq waited until the washing and cutting was out of the way before making her request. 'I'm looking for a woman named Camilla Hatton. Short white hair. Not from round here, but I think she worked in Teesside. Fifty-ish, sporty.'

Natalie shook her head. 'Not a client of mine. D'you want me to check Hairnet?'

Natalie had developed and licensed a phone-based booking

system which had been taken up by most of the salons in the country. Coupled with a refreshingly relaxed attitude towards data protection, Natalie was a valuable source of information.

She typed the name into her phone. 'No. No one by that name booking through Hairnet. Shall I put a call out to the continental sisters?'

'Please. I need to find her.'

Natalie tapped a few keys. 'Done.' She asked no further questions about the search as she started to dry Jaq's hair. 'You've got a little grey, you know.' Natalie combed through Jaq's long, thick tresses. 'Want me to do something about it next time?'

'No,' Jaq said. And then, 'It's natural.'

Natalie reached for a colour chart, but Jaq had other things on her mind. She sat up straight, pulse racing, staring at the mirror. Had she imagined it? A bearded face at the window. Fleeting, but furtive. She looked over her shoulder. The face had gone. But the space that had seemed cosy a minute ago now felt confined.

'Sorry, Nat.' Jaq sprang from the chair. 'I need to check something.'

She peered through the window. A man was moving along the street. She hadn't seen him walk past, so he was returning the way he came. He had his back to her now; all she could see was his tartan shirt and fur-lined boots. Something familiar in his gait. Was she imagining things? She shook her head. *Bolas.* Best not to take any chances.

'Can I slip out through the back?'

Natalie raised a perfectly threaded eyebrow and led the way. 'One condition.'

'Name it.'

Natalie's eyes sparkled. 'You tell me the full story next time.'

'It's a deal.'

Jaq hurried down the cobbled alley, designed for horse-drawn carts to deliver coal straight to the back kitchens of the railway terrace, scalp prickling as an icy spring breeze caught her damp hair. She pulled her coat tighter around her waist and picked up the pace. Glancing along the street as she approached the main road, she pulled back into a garage doorway just in time. *Valha-me Deus!* No doubt about it. Blackbeard. The driver who delivered the Zagrovyl order. The one who had insisted on the samples. Boris. He was waiting for someone, waiting for her. Run.

Heart pounding, she retraced her steps. Past the back door of the barber's shop, round by the telephone exchange and then along the railway track. Egglescliffe village clock struck ten, the chimes reverberating between the road and rail bridges. The Manchester airport train was ten minutes away and the Grand Central to London had already left. She slung her bag across her chest, bent low and sprinted across the railway viaduct, high above the River Tees.

On the Yarm side of the bridge, a vertical ladder took her down to the river path. She took the long way round so she could observe her flat from the opposite side of the high street.

Just as well. And even worse than she feared. The second man was easy to miss: head down, wearing a long fawn raincoat and beige flat cap, almost a study in nonentity. He walked past her flat, staring up at the windows. Once might be the general curiosity of a tourist, twice might be property-hunting . . . six times was surveillance. No doubt about it, Mr Beige was waiting for her.

Jaq sped away from her flat, taking the muddy footpath beside the Tees. The river rolled and boiled, peaty-brown water carrying tree branches from the wooded slopes of the Pennines towards the North Sea.

Blackbeard and Beige. Who sent them? Frank, of course. Maybe she'd asked too many questions. Just like Camilla? Maybe she'd better disappear before they helped her on her way.

At the crossroads, she sat on a bench and checked her bag. Car key, passport, credit cards. Everything she needed for a trip. Time to get out of Teesside.

She called Johan.

Sunday 13 March, Cumbria, England

The cottage nestled into the hillside. The thwack of an axe on wood punctuated the roar of rushing water. A stream tumbled over a cliff, through hawthorn trees and mossy rocks, skirting the garden before sliding towards the lake, a silver crescent of water far below.

Jaq parked in between Johan's trailer and a shed stacked with kayaks. As she opened the door of the Land Rover, the familiar scent of moss and fern made her smile. Johan appeared with a basket of logs, dressed only in shorts, bare-chested despite the rain. Looking serious. Looking fit. Looking fine. Her heart leapt. Boy, was she glad to see him.

Their embrace was interrupted when the front door flew open and a little boy barrelled out, followed by a black and white puppy. Johan released her and held his arms out for his son, swinging the boy in the air to his shrieks of delight. The puppy barked and ran in excited circles around them.

'You remember Ben?' Johan asked Jaq over his shoulder.

The dog or the child? Johan kissed his son. The boy, then. He'd been a sleeping infant last time. He looked more interesting now. Jaq waved a hello, but the boy turned away, burying his face into his father's chest.

The front door opened directly into a large farmhouse kitchen. Johan's wife was waiting in the doorway, a small blonde woman with a generous smile.

Jaq took the proffered mug of tea, inhaling bergamot and steam, almost spilling the hot liquid at the squeak of springs as a baby swung towards her, suspended in a harness from one of the oak beams inside, fat little legs pumping vigorously.

'Jaq,' Emma said. 'Meet Jade. The latest addition to our family.'

Jaq put down her tea and approached. 'Hi, Jade,' she said. Two bright blue eyes stared. The rosebud mouth trembled, and the baby began to wail.

Jaq took a step back, almost tripping over a cat curled up on a rug. The cacophony of noisy children and animals rang in her ears. *Santissima.* What had she been thinking of? This menagerie might be Johan's idea of perfection, but it was certainly not hers.

'Husband, *will* you put some clothes on.' Emma affected mock irritation; no one could miss the love and pride, least of all Jaq. 'We have guests.'

'Jaq's not a guest,' Johan said. 'She's my best mate.'

Best mates. Friends. That's what they were. That's what they had always been. And always would be.

Emma pursed her lips. 'And it's bath time.'

'My turn?' Johan asked. His wife nodded, raising her face for a kiss. 'Come on, kids.' He scooped them up, one under each muscled arm, and propelled them, squealing and giggling, upstairs.

'It's lovely to see you, Jaq,' Emma said.

Jaq forced a smile. 'And you.' It wasn't quite a lie, though it was Johan she had come to see. Johan she needed to talk to. When had he surrounded himself with all this extra baggage? No longer able to drop everything at a moment's notice and give her his full attention. She gazed at the stairs, willing him to come back down.

Emma chattered about her parents, who lived nearby, and Johan's, who didn't. 'How's your mum?' she asked.

The anger took Jaq by surprise. It started in the pit of her stomach, an acid drill that seared as it swirled up through her gullet, constricting her throat and stinging her eyes.

It wasn't Emma's fault for asking. Jaq had only told Johan that her mother was sick. How was Emma to know the full story?

'No change,' Jaq said. Still in a nursing home. Still bereft of her wits. Still so upset by her only surviving child's visits that Jaq rarely made the long journey.

Emma opened her mouth as if to ask another question, then pursed her lips instead. They sipped tea and the unasked questions hung between them like a net of barbed wire.

Did Emma mind Jaq being here? Johan had never made any secret of their prior relationship. Or that it was over for him by the time Emma came along. Jaq studied the face of her best friend's wife. A heart-shaped, pretty face; apple cheeks, lightly freckled, softly curling fair hair. Open blue eyes that showed no hint of guile. No edge to Emma.

'More tea?' Emma jumped up to open the fridge. 'Or wine?'

Jaq smiled. 'Guess.'

Emma filled a large glass for Jaq and poured a small one for herself. 'Don't tell Johan. He's on a health kick, and I'm still feeding Jade, but a little sip won't hurt.'

'Cheers.' Jaq sipped the cold, straw-coloured liquid. Aaah. Lime and gooseberry, tiny bubbles, sharp on the tongue, refreshing. She sat down on a bench at the kitchen table, running a hand over the rough pine surface, a finger tracing the whorls. Another sip. Better already. Come on, make an effort. 'How's business?'

'Booming.' Emma pulled up a chair opposite her, the wooden legs screeching against the flagstones. 'Johan has never been busier. Stag dos, hen parties, kids' camps in the holidays.' She raised a glass. 'But the real money is in corporate team building.'

'And your legal practice?'

'I'm on maternity leave. Still doing the pro bono work to keep my hand in.' Emma put her glass down. 'Refugee charity.'

Jaq cocked her head. 'Thanks for providing refuge.'

Emma leant forward. 'How was Slovenia?'

'Complicated.'

Emma grinned. 'Man trouble?'

'The least of the complications.' Jaq closed her eyes and an image of Karel flashed into her mind. Their goodbyes had been brief and tender. He promised to wait for her, but why would she choose to be in Slovenia without a job?

There was little opportunity for adult conversation over dinner, a noisy family affair featuring Johan's signature pasta bake, but once the children were finally in bed, the three adults, one puppy and a cat moved to the snug, gathering round a log fire.

Emma led the interrogation while Johan listened intently. Jaq told them about the delivery mix-up, the disappearing samples, her suspicion that Camilla had copied her keys, the break-in, the explosion, the inquiry, her suspension and Camilla's disappearance.

'You've contacted the police?' Emma asked.

Jaq thought back to the young detective who sat silently throughout the kangaroo court proceedings at Snow Science. 'The Slovenian police are on the case.' Her lack of confidence was audible.

Jaq met Johan's gaze. Quizzical blue eyes. A deep well she could tumble into. For the first time she was tempted to say it aloud, to crystallise the terrible theory haunting her. She dropped her gaze to the fire. Coils of glowing embers writhed on a burning log, Kekulé's Serpent in the flames. If she gave voice to her fears, she could no longer sit on the sidelines. Could she carry this burden? Better to leave it to the professionals.

'Fact of the matter is, I was responsible for the explosives store, and the explosives store blew up. And the prime suspect, Camilla, has disappeared.' It was as if the woman in turquoise salopettes had never existed. 'Unless I can prove Camilla's part in all this, it's my word against my boss's. The odds are stacked against me.' She needed space and time to think. 'Let's change the subject.'

'Music?' Johan asked.

'Please.' Jaq smiled. 'And now, I want to hear about you guys.'

Johan selected Archie Shepp. The solo tenor saxophone burst into the room, brave and bright. They stopped to listen, transported by the applause to a jazz festival in Montreux. The free-jazz cadenza sank into a mellow quintet. Warmth and peace. Good food and good friends. Music to soothe the soul. A 'Lush Life' indeed.

Johan turned down the volume, encouraging his wife to talk about the advocacy work she was doing. After some initial reluctance, Emma became animated. 'There's one guy, a photographer . . .'

Ben appeared at the door, hair tousled, half-asleep. 'Baby crying,' he said.

Emma got up. 'Sorry, Jaq, I need to go and feed Jade. Let's talk more tomorrow?' She kissed Jaq on the cheek and turned to Johan. 'Darling, not sure who is sleeping where.'

He got up. 'I'll sort it.'

Jaq stared at the fire. Through the silence came the pitter-patter of rain outside, the slap of wet leaves on the window pane, the crackling and hissing of the logs.

Johan returned with a sleeping bag and a bottle of brandy. 'Musical beds in this house. Wherever Jade or Ben go to sleep, they seem to gravitate to their mum during the night. Our bed is getting too small for four.' Did he mind? It didn't look like it. A good father.

Johan poured Jaq a glass and one for himself. 'Don't tell Emma about this,' he said. 'She can't drink while she's breastfeeding, so I'm trying to cut down as well.' He took a sip and sighed with pleasure. 'Suspicious samples, explosions, corporate impostors, it's a mystery, all right. Your life is never dull, is it, Jaq?' He threw a log onto the fire; red and orange sparks fanned out and danced up the chimney. 'So, who do you believe? This Camilla woman, or the Zagrovyl people?'

'I don't know what to think. Camilla came across as smart and plausible. Frank is a dick.' Jaq sipped the firewater. 'Camilla was hiding something, and yet I would trust her further than I could ever throw Frank Good. He claimed not to know her, but he was lying.'

'You think Frank got rid of her?'

'I don't know what to believe. Zagrovyl is a publicly listed company, FTSE 100. Their shareholders are unlikely to endorse cold-blooded murder.'

Johan wrinkled his nose, unconvinced. 'What do you think they were transporting? Drugs?'

Jaq took a long swig. 'Worse than that.'

Johan got to his feet. When he came back, he was carrying a folder.

'Emma's pro bono case,' he said. 'Feeling strong?'

Jaq nodded.

Inside the folder were large-format black-and-white photographs. The first picture showed the corpse of a child aged three or four, about the same age as Ben, lying on a concrete floor in a foetal position, face turned away, knees drawn up to the chest, arms wrapped around thin ankles.

The second picture revealed that the child was just one of a family of six who perished together. Behind the boy, on a divan bed, lay a stick-thin old woman. Beside the boy, a young mother clutched a lifeless infant. Beyond the door of the one-roomed house, two men had just made it to the road outside before they collapsed and died.

The third picture showed a whole village exterminated. Hundreds of people lying on the ground. Animals, too: dogs, cats, goats and birds that had plummeted from the poisoned sky. No gunshot wounds, no blood, no shrapnel, no physical damage. Buildings intact. Living beings felled. Eerie. Heartbreaking.

The last picture was of the first subject again, the little boy, photographed from a different angle. Bile rose to her throat as her eyes moved to his face. This child had suffered. Oh God, how he had suffered: head thrown back at an impossible, neck-breaking angle; clouded, sightless eyes open and protruding from shock and terror; mouth wide open, jaw almost unhinged in his last, terrible scream.

'Sarin gas,' Johan said. 'One of the few attacks that was properly documented immediately afterwards.'

Jaq wiped her eyes. 'How did you know what I feared?'

'I've known you a long time, Jaq.' He took her hand. 'I know when you are not telling me everything.'

'But I don't know anything, not for sure.'

'The stuff you sampled, the stuff that went missing, you suspect a link to chemical weapons?'

Jaq nodded. 'I can't rule out that possibility, but—'

'I've witnessed how chemical warfare affects people.' Johan's mouth tightened. 'Even if it's only a hunch, you need to escalate.'

'I don't have a shred of evidence.'

He put an arm around her. 'Jaq, your gut feel is worth more than a million samples.'

Ben reappeared at the door, his father's chaperone.

Jaq withdrew into the sofa and hugged herself. 'You get back to your family,' she said. 'I'm going to sit by the fire and think a little longer.'

'You okay with a sleeping bag?' He pointed to the futon.

Jaq nodded.

'Put the fireguard on when you turn in.' Johan paused in front of the fire, silhouetted by the flames. 'Goodnight, Jaq, sure you don't need anything?'

Anything? I need everything. And nothing. 'Goodnight, Johan, I'm fine.' She reached for the bottle of brandy and turned back to the fire.

Monday 14 March, Cumbria, England

Patches of snow glistened on the high hills, but the rain swept the rest down in ribbons. The ribbons formed streams, the streams combined into rivers and the rivers rushed down towards the lake.

Even after the phone-tracking signal was lost, it didn't take Boris long to find the right place, the battered green Land Rover instantly recognisable beside a rack of kayaks in an open-sided shed. Boris drove slowly, observing the large garden with swing and slide, the lake in front, cliff and stream behind. He cruised past, accelerated round a sharp bend and found the National Trust car park beyond. Pulling on his brand-new green cagoule, he folded the Ordnance Survey map into the pouch and hung it around his neck, along with the binoculars. Master of disguise, Boris walked back to the ridge to get a better view. Whitewashed stone. Thick walls and small windows. Old, the English were obsessed with old. Old cars, old houses. Idiots.

The farmhouse door opened. A man came out and began to load up the trailer with kayaks. The next person to appear was Silver. She helped the man connect the trailer to the jeep, although judging by the muscles on him, he needed no assistance.

A short blonde woman emerged with a babe in arms and a little boy scampering around her feet. Boris didn't like children. He had fathered none himself, at least not as far as he knew. No brothers or sisters, and he kept his distance from his mother's extended family in Slovakia. Gypsies. Filthy animals. In his limited experience, children were drippy, noisy and unpredictable. You scowled at them and they laughed, you grinned at them and they screamed. And went on screaming. Boris didn't like children at all when they were awake.

But the little one was sleeping. She looked quite appealing in that state, he had to admit it. The little rosebud lips, pouting and sucking on an imaginary nipple. A cloud of fair hair, so light and insubstantial it seemed to hover around her like a halo. The soft skin would be velvet to the touch.

Silver looked away as the man kissed Blondie on the lips, then boy and baby on their foreheads. The man didn't even acknowledge Silver in the presence of his family. Such studied avoidance. Interesting. *Watch out, Blondie, you have competition. A big black cuckoo in that fair nest.* The man got into the Land Rover and drove off. Good. He looked inconveniently fit. One less to worry about.

How to do it? How to silence the *kurva* who thought she could outsmart him? How to do it quickly and silently? Without getting caught? This was England. Normally he wouldn't risk anything here. But she'd left him no choice. He had to clean up before The Spider found out.

He'd watch and wait.

Improvise.

Monday 14 March, Teesside, England

Frank was leaving work for a lunchtime meeting when his phone rang. He stepped under the gatehouse awning and checked the number.

'I have good news.' The Spider spoke softly.

'About bloody time.'

'All is well. The numbers do not lie.'

'And the missing materials?'

'Incompetent management. Computer system error.' The Spider tutted. 'No further cause for concern now you have sacked crazy Ivan and the bad apples.' A laugh. 'The books balance.' The Spider stretched the last syllable into a long hiss.

Frank played his trump card. 'So, who are SLYV and what are they paying for?'

A splutter, then silence.

'Look, you hopeless piece of shit,' Frank said. 'I'm not stupid. I can read management accounts. Don't fucking lie to me. Next time you call me, you'd better have something useful or I'll have you struck off.' Take that, you useless bastard. Frank cut the call.

As Frank waited for Bill in Café Lilly, he ran through the Brandenburg Concerto No. 5. The show-off solo cadenza. Frank had an excellent musical memory. For contemporary music, he sometimes had to pause and refer to a recording or the score. But for baroque music, for J.S. Bach, he was note-perfect. When he sat stony-faced in meetings, he was enjoying the sublime music in his head, usually more interesting than listening to other people.

Bill was more interesting than most. Frank saluted as he entered the restaurant.

'Well?' Frank put his hands behind his head and spread his legs wide. 'What have you got for me?'

'Unusual story, this one.'

The waitress came to take their order.

Frank ordered the fish special for both of them and waved her away. 'So, tell me.'

Bill laid his iPad on the table. 'Born Maria Ines Jaqueline Ribeiro da Silva, Luanda, Angola. Angolan mother of Anglo-Portuguese descent, Russian father.' He flicked to the birth certificate.

Older than she looked.

'One brother, died young. Family left Angola during the civil war. School in Moscow, Lisbon and then Middlesbrough. Parents separated. Changed her name to Silver when she moved to the North of England. Honours degree in chemical engineering from Newcastle University, master's from Edinburgh and PhD at Teesside University.'

An academic.

The fish arrived, skate wings with a black butter and caper sauce. Frank ate while Bill talked. No point letting both meals get cold.

'Father dead. Mother in an institution.'

'Institution?'

'Mental health problems.'

A family history of mental illness. Could be useful.

'Worked for ICI. Closely connected with Teesside University, both for her PhD and then as visiting industrial lecturer. Married Gregor Coutant. Separated. Divorce not yet finalised.'

Broken family. Unstable relationships. Perfect, he could use that.

'She left ICI shortly after Zagrovyl took over. Cited in a private prosecution over the deaths at Seal Sands. Moved to Slovenia and took up a job with Snow Science as their explosives expert.'

Snow Science again. Explosives in the hands of a mentally unbalanced criminal. Lovely.

'Quite the party animal. Observed stepping out with various

young men in Kranjskabel. Most recently, ski instructor Karel Žižek.'

Juicy. 'What do you have on the latest boyfriend?'

'Squeaky-clean. Suspicious. I'll dig a little further.'

Frank polished off the last morsel of fish and pushed his plate away. Snow Science, SLYV – what was Silver's part in all this?

'Where is she now?'

Bill consulted his notes. 'Right now, Silver is staying with an ex-boyfriend in Cumbria.'

Frank banged his fist on the table. 'So what the fuck are you doing here?'

Tuesday 15 March, Cumbria, England

Boris had a plan.

The lake stretched out below him, a shimmering sheet of silver. At the end of a rocky promontory, out of sight of the farmhouse, a little rowing boat bobbed up and down, its mooring rope tied to a tree. Beyond it lay a hidden beach, shielded from the lakeside path by a copse of trees: silver birch, twisted hawthorn and gnarled oak. Barely a breath of wind, the water lay still and sparkling, at its deepest where the pebble beach sloped steeply and then fell away to blackness. If Silver would come down close to the water, the dangerous water, he could make the next step look like an accident.

The head of the household had already left for work, leaving only Blondie and two babies between him and Silver. Easy.

The farmhouse door flew open. Silver emerged in vest, shorts and trainers.

'Have fun!' A woman's voice floated out from the farmhouse kitchen.

'Back in thirty minutes!' Silver closed the door.

He looked away as she did her stretches, his throat drier than a desert as the bounding pulse in his neck intensified. The straining breasts. The taut buttocks. The long legs. Disgusting. Christ Almighty, had she no shame?

She set off with a long stride. Heading up the hill, not down to the lakeshore. Away from the freezing water. Damn.

What would bring her down?

The door opened again. The little boy skipped out, followed by Blondie. He held a toy boat in his hand and made outboard motor noises as he scurried around the lawn.

'Oi! Stay where I can see you,' his mother admonished. 'Don't go near the water.'

Oh, but that is exactly where you are going.

Blondie hefted a wicker basket onto her hip and headed for the woodshed. A shed with a padlock on the outside. Perfect. Couldn't be easier. Child's play.

Boris crossed himself and asked forgiveness for what he was about to do.

Child's play.

No, Boris didn't like children; they brought out the worst in him.

When Blondie was fully inside the shed, Boris snuck round the side and closed the door.

'Ben?' she shouted. 'Is that you?'

Boris threaded the heavy padlock between the door loop and the metal clasp. It snapped shut with a satisfying clunk.

'Mummy?'

The little boy came running from the other side of the garden, unaware of the intruder until he reached the locked door of the shed. Boris pocketed the key and grabbed the boy by the neck. The child stared at him wide-eyed, shocked into silence, his little arms and legs windmilling uselessly as Boris raised him up and clapped a hand over his mouth.

He inspected the child. Not as appealing as his sister. Pity. Fewer entertainment options while they waited. But better bait. This one would scream his head off, shout directions and bring Jaq running straight into the trap. A ghastly accident in a freezing lake.

He whispered in the boy's ear. 'Let's leave your mother in the woodshed.' He slung the boy under his arm. 'While we play. You, me,' he started the descent towards the icy water, 'and Silver.'

The path down to the lake was steep and uneven. Boris paused at a grassy platform halfway down. The little boy wriggled under his arm, punching and scratching, kicking and biting, throwing him off balance. An amusing tussle when there was no hurry. Warm little boy. Soft little boy. Weak as a kitten. Powerless, just like Blacky, Whitey, Spotty, Stripy, Softy.

Boris yanked the boy up to face him, one big hand spanning both tiny armpits.

'Stop fighting,' he growled. 'Or I might have to hurt you.'

'I don't care.' The boy spat in his face. 'Let my mummy out of the shed.'

Boris wiped the tiny drop of warm saliva from his cheek and smeared it over the boy's face, running a fat finger down from the child's temple over his button nose to the trembling lips. Soft lips. He forced them open with his finger. 'Do you love your mummy?'

'Yes.' No more than a whisper.

'And your little sister?'

'Yes.' Louder this time. Defiant.

'Do you want me to hurt them instead?'

The boy began to cry.

'Then shut up.'

Until I make you shout.

Boris reached the lakeshore, the child under his arm quiet now, apart from the sobbing. He bounded over the rocks to the mooring point, unhooked the rope and walked the boat to his chosen place under the cliff. The perfect spot, only one way in, down a narrow, twisting path. Where he'd be waiting. And for Silver, only one way out. Underwater. Until she stopped breathing.

How much time left? Silver must have turned back by now. Boris pulled at the line. Wood juddered against pebbles as the hull mounted the beach. He reached in and removed the oars, tossing them into the water.

'I'm going to set you free now.'

The little boy scowled at him, suspicion fighting with relief.

'What is your name, little boy?'

'Ben.'

'Get in the boat, Ben.'

Ben shook his head. 'I can't swim. I'm not allowed—'

'I said, get in the boat.'

Ben clambered awkwardly into the flimsy wooden craft and began to root around under the seat.

'What are you doing, Ben?'

Ben emerged with a yellow life jacket and held it aloft. 'Daddy says—'

Boris snatched it from him and tossed it into the water where it bobbed beside the floating oars.

'Whatever you hear next, whatever you see next, don't tell anyone.' He brought his face close. 'If you tell anyone about me, I'll come back for you.' He threw the rope into the water. 'If you tell anyone what happened, I'll be back to hurt you. And I'll hurt your baby sister, too.' Was the boy too young to keep a secret? Better not take the chance. He leant into the scuppers and pulled out the bung. 'I was never here. You went out in the boat all by yourself.' The leaky boat. 'Now, off you go.' He kicked the boat off the beach and gave it a shove. 'Shout for Silver to come. She's your only hope now.' The boat glided into deep water. Silver would be able to see it from the hillside above. 'You haven't got long. Shout for Silver if you want to stay alive.'

'Mummy!' Ben screamed at the top of his voice.

The boat rocked as the little boy stood up. *Careful now. Don't spoil things by drowning too soon. All in good time.*

The boy was wailing like a banshee as the boat took on water. A noise in the trees. Footsteps. Someone running down the path.

Silver was coming.

And Boris was ready.

Tuesday 15 March, Cumbria, England

Sunlight dappled the mountain, the sphagnum moss springy underfoot, little jets of water squirting around her trainers, throwing up tiny rainbows as she passed.

Fell running required total concentration. On an indoor track you could let your thoughts wander, free mind from body as the muscles contracted and relaxed in a regular rhythm. But the mountain was littered with hazards: flakes of smooth grey-green slate that slid over each other; narrow ravines of fast, glittering water that cut deep into the turf; a sudden squelch of bog cotton and frogspawn among twisting hummocks of coarse grass and moss; and sharp, loose rocks. One false step could spell a twisted ankle, knee sprain, bursitis. A fall could fracture a rib, a collarbone, break an arm. Or worse, concussion with no one for miles except the bleating sheep.

Beynnn, they bleated. *Memmmmy* . . .

At the first false summit, she stopped to catch her breath. What now? Carry on a bit longer, or head back to help Emma? Freedom or responsibility?

Jaq chose freedom. Every time.

She pushed on, dropping down out of sight of the lake into a mountain valley, getting into her stride, pushing herself hard, welcoming the burning sensation in her muscles.

'Mummy!' The scream carried across the water, up the mountain and into the valley, echoing between the cliffs.

That bleating sheep again, high-pitched but this time a word. 'Ben!'

Jaq raced to the next ridge and looked back down at the lake. A tiny boat rocked in deep water, just beyond a beach hidden by the cliff. A life jacket, a yellow dot bobbing on the blue water, floated away from the boat, trapped by a wooden cross formed from the abandoned oars. Why was the boat rocking on such still water? *Diabos me levem!* Was that a tiny figure moving inside? Surely not?

Jaq ran straight down the mountain, ignoring the path, jumping over rocks, squelching through bog, leaping across streams, sliding over scree, cutting through every obstacle, running a straight path to the lake. Feet flying, lungs burning, arms pumping, running faster than she had ever run before.

'BEN!' No mistaking it this time. That was no sheep bleating, it was Emma shouting. Muffled, distorted. Where the hell was she? 'JAQ, HELP!'

Ó meu Deus. Faster.

Getting closer now. No doubt about it. Ben alone on the water. On the deepest part of the lake. Little Ben who couldn't swim yet. In a boat without a life jacket. Without oars. A boat that was tilting dangerously.

'Ben!' she shouted. 'Stay still, I'm coming.'

'Help!'

The boat was listing. *Credo!* The boat was sinking.

'Emma, where are you?'

The bleating came from the farmhouse garden. 'Jaq, I'm . . . I can't . . .' The words lost behind some barrier. A baby crying in the farmhouse. Jade. Why was she not with her mother? No time to deviate. Time to calculate. Fastest way down? Ben didn't have much time. Only one option.

Jaq sprinted to the cliff above the hidden beach. No time to kick off her trainers. No time to stop and measure the height she would fall, the width she had to jump, the depth of the water if she cleared the beach. No time to hesitate, assess the risk, take the slower path down. No time to lose. Now or never.

Running at full speed towards the overhang, Jaq launched herself in a running jump, spinning forward as she tumbled through the air.

The thought experiment was instant, calculation without words. Acceleration due to gravity: after one second her vertical speed would be 9.8 metres per second, after two, 19.6 metres per second, after three, 29.4 metres per second. Terminal velocity 50 metres per second, but she'd never reach that – the cliff couldn't be much more than 30 metres high. A 2.5-second fall at best. Horizontal speed at launch? Sprinting speed. World record on a track: 100 metres in under 10 seconds. Ten metres per second tops. The clifftop was not track. At half that speed, and assuming no resistance from the air, she would continue moving at running speed for 2.5 seconds. At 5 metres per second it gave her 12 metres horizontal travel before she hit the ground. It's not the fall that kills you, it's the sudden stop. Was the beach too wide? Could she make it to the water?

Oh, Ben. *Ben.*

Jaq hit the water in a shallow dive, her fingertips stretched to cut the water first, scraping her legs on the pebbles as they followed, surfacing and gasping at the shock of the cold water. Alive. Exhilarated.

'Help!'

Hurry! Fast crawl towards the boat, the trainers swelling with water and slowing her down. No time to untie them. Keep going. Adrenaline pumping. She grabbed the life jacket as she passed it, hooked it over her shoulder, heading for the boat.

Suddenly nowhere to be seen.

'Ben!' she screamed.

A little tousle of fair hair bobbed up ahead. *Graças a Deus.* The boat was under the water, but Ben was still clutching the prow, shivering with cold and terror.

'Good boy. Hold on. I'm coming. I've got you.'

He threw out his arms, grabbing her round the neck, choking her, pulling her down.

Gasp. Breathe. Tread water. Kick, kick. Breathe. Gasp. Disentangle. Gently now. Kick, kick. Gasp. Breathe. He may be small, but he could drown you. Prise his cold hands away. Not so gently now. Yank him loose. Life jacket over his head, on his back, on your back. Swim for shore. Kick. Kick. Breathe.

Christ, how had the boat travelled so far? The beach didn't seem be getting any closer. And Ben had stopped struggling. Stopped moving altogether.

'Ben. Come on, kiddo. Swim with me. I need your help!'

How to keep him conscious, keep him warm? His cotton T-shirt billowed through the life jacket, too loose to preserve any heat, his denim jeans sodden and heavy, his little shoes lost in the lake. She supported his tiny body, the soft skin so cold against hers. What did he weigh? Less than fifteen kilos. A metre or so tall. How long had he been in the water? He would lose heat fast. Surface area to volume ratio so much greater than hers. And she was moving, straining every muscle to get him to the shore, to get him to safety, to get him warm, as fast as humanly possible.

She heard the cry from the middle of the lake before she spotted the kayak.

'Jaq!'

The red ripper sliced through the water, the single paddle flying so fast that it formed a continuous halo around the kayak. Johan. Thank God.

'Ben! Is he OK?' His deep voice boomed across the lake.

What to tell him? The child was cold, so cold. Still, so still. She couldn't even stop to check his breathing. And what could she do? CPR impossible in the water. His father was still too far away to help them. Swim. Swim. Don't answer him. Save your breath. Swim. Swim.

The flashing blue light ahead gave her renewed energy. Oh God, let it be an ambulance. Please let it be an ambulance. Please let there be a paramedic able to take this tiny, cold body and breathe life back into it. Getting closer now, finding her feet on the pebbles, thankful for the shoes she had cursed in the water, sodden trainers that now let her fly across the submerged pebbles.

A policeman appeared on the beach with a hacksaw and axe in his hands, followed by Emma.

'Ben!' she screamed.

The frozen puppet stirred in Jaq's arms.

'Mummy!'

Oh God, he was alive. Ben was alive!

As Emma enveloped her son in a warm embrace, Jaq collapsed onto the stone beach and looked back at the lake.

To see Johan in his red kayak dragging a body though the water.

Tuesday 15 March, Cumbria, England

The police car crunched over gravel, heading back towards Ulverston. Jaq glanced at the lake, now crimson beneath the setting sun, and the anger boiled up inside her. Volcanic, incandescent, flaring anger.

Anger with herself. How could she have been so selfish? Running to Johan because men were following her. What had she been thinking? Had she considered the consequences? For Johan, yes. Johan was ex-army, he could handle anything. But how could she have ignored the danger she was bringing to his wife, sweet Emma, his children, Ben and Jade? What sort of monster had she become? Disgorging her own trouble, vomiting it onto innocent children.

Anger with the man sent to kill her. She recognised him. The man spying on her in Yarm must have followed her to Cumbria. He was leaving now, in an ambulance with a police escort, straight to the morgue.

Anger with the police. She was the last to be interviewed and she told them everything: the explosion at Snow Science, Camilla Hatton's disappearance, Frank Good's lies, his henchmen following her, Zagrovyl's curious connection with Snow Science. As if a dam had burst, the torrent of words, suspicions, certainties flowed freely. Better than tears, better than anger, it was such a relief to unburden herself that she barely registered the police reaction. All that mattered was that they left with a promise to contact Interpol and OPCW – the Organisation for the Prohibition of Chemical Weapons.

The gate clunked closed. Johan stood in the garden, his back to her, staring at the sunset. Her anger dissolved into shame. Despite the dry clothes, the hot shower, the sweet tea, she welcomed the shivering that racked her body, the throbbing ache in her temples and the gripping nausea in her stomach; they all gave physical form to her guilt.

She shrugged on her coat and went out to him.

'Emma's taken the children to her mother's.'

'How's Ben?'

'Temperature back to normal. A few bruises. Otherwise physically unharmed.'

'And Jade?'

'She was asleep for most of the drama, safe in her cot inside the farmhouse.'

'How did you know to come back?' Her jaw ached with the effort of keeping her teeth from chattering.

'Emma called me from the woodshed. And the police. She had her mobile with her. I was teaching at the other end of the lake – I took the quickest way back.'

'Will she ever forgive me?'

Johan put an arm round her shoulders. 'You saved our son, Jaq.'

Jaq ducked, twisting away from him. 'Don't.' She put her head in her hands. 'Coming here was a mistake. I put you all in danger. I'm so sorry.'

'Did you kill him, Jaq?' Johan asked. 'That man in the lake. I wouldn't blame you if you did.'

'No!' How could he think that of her?

'Then what happened?'

'What did Emma see?'

'Nothing. She went to get wood for the fire and next thing she knows she's locked in the shed and Ben is crying outside. She's sure she heard a man's voice, sure it was the man in the lake.'

'What did Ben say?'

'He won't talk about it. He's still in shock. The police insist he must have locked his mum in the shed as a prank and gone on a solo boating adventure that went horribly wrong.'

'What about the body?'

'The police theory is that a passer-by tried to rescue Ben and drowned. Only the autopsy will tell.'

'And you, do you buy the police theory?'

'Of course not. It's ridiculous. I told them so. Ben can be a tearaway, but he obeys his mother. And he understands boats and water, has a healthy respect for danger. Christ, Jaq, he's four years old.' Johan's voice broke. 'Ben can barely reach the padlock on the woodshed, far less force it shut.' He blinked hard. 'Did you see anything?'

Jaq shook her head. 'I went for a run. Emma, Ben and Jade were all in the farmhouse. Next thing I know, there's a boat sinking in the middle of the lake with Ben in it.' She shivered. 'The man who drowned, I recognised him. I know he was sent to kill me. Make it look like an accident. But something went wrong, and Ben raised the alarm. He's a hero.'

'I'll tell him that.' Johan jangled his car keys. 'I'm going to my in-laws now.'

Of course. He wanted to be with his family. Time to move on. 'I'm heading off, too.'

'Back to Teesside?' Johan didn't try to dissuade her. 'Whatever happened today, the man who followed you is dead. The police are on your side. You're safe now.'

Jaq wished she could be so sure.

Wednesday 16 March, Teesside, England

The rain pelted the police car, the north wind driving the droplets at a 45-degree angle, soaking the passenger as the uniformed officer opened the door. Detective Inspector Dias unfurled a golf umbrella, coloured triangles of green and purple plastic straining against the light metal frame, and hurried towards the sliding doors of the Zagrovyl headquarters.

Frank Good's assistant, sharp cheekbones and well-cut suit, greeted the policemen and whisked them past the blank-faced receptionist, away from the public area with its black leather sofas and revolving trophy cabinet, to the private conference suite where the Zagrovyl director of European operations was waiting.

Introductions were made, coffee dispensed – black for DI Dias, no sugar, thank you, white for Sergeant Prosser, two lumps, please – before the four men sat down, the policemen facing the wall, the Zagrovyl representatives facing the glass door.

'Now then, Officer, you said it was urgent.' Frank placed his hands on the table. 'How can I help you?'

DI Dias slid a photo across the polished surface. 'Do you know this man?'

Frank considered his options. Nothing to be gained from concealment; the police must already know.

'A private investigator. Bill Sharp.'

'Working for you?'

'He does occasional odd jobs for me, yes.' Frank sat back and opened his legs. 'Why do you ask?'

'Found dead yesterday morning. Drowned in Lake Coniston.'

'I see.' Frank scowled and then corrected himself, making an effort to appear to care about something other than the dammed inconvenience. 'Poor Bill, how dreadful. What happened?'

'Too early to say. Do you know a Dr Jaqueline Silver?'

Maybe she was dead too; that would solve a few problems.

'Not well. I met her only once. She came to this office last week. Friday, wasn't it . . .?' He looked over at his assistant for corroboration.

'I can check the visitors' book, sir.'

The useless mannequin, all full lips and pointy shoes, remained seated. Nicola had advised against him at interview, which was precisely why Frank had hired the sharp-suited lad. But what fucking use was an assistant if they couldn't back you up when you needed them?

'You do just that,' Frank said. When the fashion lump didn't move, he waved his hand. 'Off you go.' He waited until the moron got the message before returning his attention to the detective inspector. 'Why do you ask?'

'She has made some rather serious accusations.'

So, she wasn't dead, but still alive and causing trouble. What was her part in Bill's untimely demise? Frank narrowed his eyes. 'Can you elaborate?'

'All in good time.' The DI coughed and pulled out a notebook. 'Perhaps you could tell us what . . .' he checked his notebook, '*odd jobs* . . . the late Mr Sharp was doing for you?'

Frank took out a white cotton handkerchief and blew his nose. 'Are you married, Detective Inspector?'

'Divorced.'

A man after his own heart. 'This is a little embarrassing, Detective Inspector.' Frank put his handkerchief away. 'I first hired Bill at a troubled time in my marriage. I suspected my wife of . . .' He dropped his eyes and let the silence imply infidelity. He certainly wasn't going to elaborate on the real trouble with garrulous mistresses which led his first wife to sue for divorce.

The DI shook his head. 'Let's focus on the present, shall we, sir?'

His tone was one of manly understanding. 'Why was Mr Sharp following Jaqueline Silver?'

There was no point in denying it. 'I asked Bill to keep an eye on her and warn me if she came back. I was concerned as to what she might do, might be capable of.' He let that land. 'Not for myself, you understand. For my family.' Which family? He had a choice of three. 'Just until I could get a restraining order in place.' Which, now he thought of it, merited a call to his lawyer the moment this interview finished.

'What caused this concern?'

Frank met his eyes. 'You are aware of her history?'

'Whose history, sir?'

'Jaqueline Silver. She was accused of manslaughter – did you know?'

The DI raised an eyebrow. So, he hadn't done his homework. Good.

'Multiple deaths. But she got off on a technicality.' Frank raised his eyebrows in exaggerated exasperation. 'The families of her victims never gave up – they brought a private prosecution. Still ongoing, still unresolved.'

DI Dias whispered something to his sergeant, who excused himself and left the room. Two down, one to go.

'There is a history of mental illness in her family. Zagrovyl tried to help, but in the end, we had to fire her.' Frank shook his head in a display of sorrow. 'Before my time, but she took it badly. Very badly. Still very bitter. Towards Zagrovyl. Towards me.'

'Why towards you, sir?'

Frank puffed up his chest. 'I guess I am the public face of Zagrovyl.'

'And how did she manifest this . . . bitterness?'

'Burst in here. Threatened me.'

'Threatened you, sir? About what?'

'God knows, she wasn't exactly rational.' Frank groaned aloud. 'Oh, poor Bill. I will never forgive myself if . . .' He slapped his forehead as if struck by a sudden thought. 'Inspector, is there any chance Jaqueline Silver knew that Bill was following her? Confronted him? There is no knowing how she would react if . . .'

Frank watched the light bulb go on behind the inspector's eyes.

'I'm not at liberty to say, sir. Can you be more specific about the threats?'

Frank joined his hands and cracked his knuckles. 'She was rambling about some woman she claimed had worked here.'

'Who?'

Careful. 'I don't recall the name. My assistant may remember.' Frank gestured at Cheekbones, hovering outside the glass door. 'Anyway, not someone currently employed by Zagrovyl.' Oh, you silver-tongued wordsmith, you. 'I regret to say that Jaqueline Silver is a dangerous fantasist, an unhinged troublemaker. She made a terrible scene. I had to call security.'

The slack-witted assistant burst in, holding up the visitors' book, all puppyish enthusiasm and eagerness to please.

'You were right, sir.' Cheekbones pushed the book across the table to the detective. 'Friday 11 March. Jaqueline Silver came here looking for a Dr Hatton.'

Good boy. The art of deception is . . . timing. 'She made rather a scene, didn't she?' What was the boy's name? Tarquin? Valentino?

Cheekbones didn't miss a beat. 'She wasn't happy. Not happy at all. But what could I do? Dr Camilla Hatton doesn't work for Zagrovyl. I had to call Mr Good.'

The DI made a note, his pen scratching against the paper. He looked up. 'Did you hear Jaqueline Silver threaten Mr Good?'

The sergeant knocked on the glass door, phone in hand. DI Dias swivelled round and nodded him in.

Cheekbones glanced at Frank. Two sets of blue eyes sparked as they met. A query, a response, a flash of understanding passed between them, all in a millisecond.

'Oh, yes,' Cheekbones said. 'The most terrible scene. Mr Good had to call security.'

Cheekbones might be a candidate for the fast track after all.

The sergeant entered and sat down next to DI Dias. He whispered something in his boss's ear and passed him a slim mobile phone. The tabloid banner glowed from the screen. The DI scrolled down to the pictures of Jaqueline Silver and the victims at Seal Sands. He sighed and snapped his notebook shut.

'I'm sorry to have wasted your time, sir.'

'Not at all, Inspector. Only too glad to be of help.' Make all the allegations you want, Jaqueline. No one is going to listen to you now.

Outside it was still raining. Frank watched as DI Dias opened his umbrella. As the policemen crossed the car park, a gust of wind caught it, inverting the metal skeleton and tearing the coloured nylon to shreds.

Jaqueline Silver: licensed to handle explosives. Already under suspicion for manslaughter, now a murderer. Well, who would have thought?

PART III: RONDO
SLOVENIA

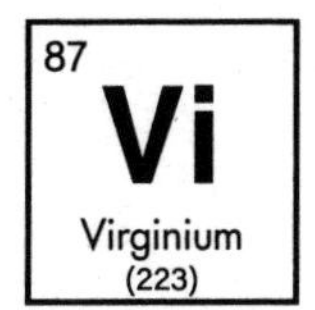

Thursday 17 March, Ljubljana, Slovenia

Jaq left England before the Cumbrian police could detain her.

The warning came just in time. The police had been back to talk to Ben and Emma. Instead of contacting the Chemical Weapons Inspectorate as promised, the police first interviewed Frank Good. He'd deflected all the accusations and raked up some muck of his own: drawn police attention to Jaq's history; accused her of threatening and harassing him; and implied that if they were looking for Bill Sharp's murderer, they should look no further than Dr Jaqueline Silver.

The new police theory, Johan informed her, was that Jaq had attacked the man following her, locked Emma in the shed and staged a fake rescue of Ben from the lake to cover up the crime.

Jaq wasn't sure what hurt more. The absurd allegation or the fact that it was Johan who phoned. Emma hadn't spoken to her directly since Ben's near drowning.

Now was not the time to dwell on a damaged friendship. Unless she moved quickly, the Cumbrian police might stop her. Detective Wilem Y'Ispe wanted to talk to her; some forensic results had come in. Time to get back to Slovenia.

The plane crossed the mountains and descended towards the flat fields where Ljubljana nestled in spring sunlight. Only a week since she had left, and already the trees were bursting with leaf and blossom.

The Specials, *Specialna Enota Policije*, operated from an old building in the centre of Ljubljana, pink stucco with red-tiled roof and ornate wrought-iron balconies looking over the town, the castle rising up behind them.

She hesitated outside. With the English police, she'd said too much too soon. When Jaq accused the man in the lake of having been sent to kill her, the detective raised an eyebrow. When she implicated Zagrovyl, he suggested she get some rest. When she demanded the police involve OPCW – the Organisation for the Prohibition of Chemical Weapons – a sergeant made her a cup of tea, told her she'd had a nasty shock and offered victim support counselling. She got angry, blurted out her worst suspicions, named Frank Good, Camilla Hatton, Laurent Visquel, made the mistake of telling the police how to do their job.

And now they planned to arrest her.

Back in Slovenia, she had a chance to start again, to avoid making the same mistake twice. This time she would stay calm. She would listen and find out what the Specials knew, identify the gaps in their knowledge, let them make their own discoveries. Every successful woman needs to know when to let a man take the credit.

A sentry saluted and opened the wrought-iron gate which led into a cobbled courtyard. A uniformed officer guided Jaq into the main building and opened a heavy oak door. Detective Wilem Y'Ispe stood at a French window looking out on to the river, his thin frame silhouetted in the afternoon light. Sunbeams danced around his fair wispy hair, but as she moved towards him, the light flickered and faded, *ignis fatuus*, will-o'-the-wisp.

'Dr Silver.' He extended a hand. 'Thank you for coming.'

'Glad to help.'

The office of Detective Wilem Y'Ispe, Will-O'-the-Wisp as she now thought of him, was large with high ceilings, in need of redecoration and sparsely furnished. Underneath the stained ceiling and peeling wallpaper stood a desk with the computer almost obscured by piles of paper, more folders on top of a row of grey steel filing cabinets and a circular table with assorted chairs. He pulled out a seat for Jaq, retrieved a thick file from his desk and sat beside her at the circular table.

'I have the forensics report for you.' He opened the file and extracted a white envelope. 'This is an unofficial copy – we're waiting for Chief Inspector Goran Trubor to approve release.'

Inside were several closely typed pages. She began to read with a growing sense of unease, frowning as she struggled to make sense of the unexpected chemical analysis in the summary.

Will-O'-the-Wisp put a hand on the report before she could turn the page to the detail.

'Before you read on, Dr Silver, I have a question. Who was in the warehouse at the time of the explosion?'

'Fortunately, no one.'

He met her gaze. Green eyes.

A flicker of doubt, a spark of alarm. She crossed her arms, suppressing the flames of fear. 'Is there something you're not telling me?'

He opened his folder again and laid a series of photographs on the table. It took her a moment to realise what she was looking at. She had seen the same scene in the explosives warehouse through the smoke and ash, but not in this detail.

Patrice, the relief security guard, had assured her there was no one working at Snow Science that Saturday morning. Her brain had taken the information and found another interpretation for the gooey mess. The photographs showed in graphic detail what happens when a soft body gets in the way of a violent chemical explosion. All that was left was a shard of jawbone with a few teeth still attached, a mutilated, severed hand, the remains of an eyeball that had popped from its socket and rolled away from the fire. What she had interpreted as melted red plastic, white electrical trunking and yellow insulation had been blood and bone and flesh.

The weariness started in her fingertips, a trembling, tingling sensation that fluttered from knuckle to knuckle, surging into the canyon of wrist and spreading up her arms, bringing a dead weight to her drooping shoulders. She tried to stand up, but a sudden fatigue swept over her body, an icy wash seeping down her spine;

her hips sagged, knees shook, ankles wobbled and she collapsed back onto the chair.

Will-O'-the-Wisp handed her a glass of water. 'Do you need a break?'

'No.' She emptied the glass. 'Sorry . . .'

'Better than indifference.' The weariness in his voice made her wonder what he had seen to make him so jaded so young.

'Our forensics team have started to recover the remains,' he continued. 'Nothing has been established yet, and we won't make this public until the press conference. But we have a missing person to identify.'

A body, or what was left of one. A jigsaw to put back together again. Someone, a human being, had died. Another one. An image of the corpse in the lake flashed into her mind and she shivered.

'So, Dr Silver, who is the victim?'

'I don't know.' Jaq bit her lip. All the employees and contractors were accounted for, no security guards or cleaners missing. It could only be someone who had no business being there, someone who broke in. Twice.

'But you have a suspicion?'

Should she tell him? About Camilla Hatton, Laurent Visquel, Frank Good, Zagrovyl? Or keep quiet and trust him to do his job? Ben's face flashed into her mind. Superimposed on the corpse of the child, the victim of the Sarin gas attack. She had to see this through. Somehow, she had to convince Will-O'-the-Wisp to act, lead him right back to where this whole mess began.

'It all started with the samples.'

'I'm listening.'

She took a deep breath. 'Zagrovyl, our supplier, made a mistake with a delivery. There was something fishy about the material.'

'Fishy, as in . . .?'

'Odd.' *Too soon to be more specific.* 'Bear with me for a moment.'

Will-O'-the-Wisp nodded.

'I took samples.'

'The ... odd ... material you sampled, what happened to it?'

'Zagrovyl picked it up, replaced it.'

'And the samples?'

'Before I could analyse them my boss destroyed them.'

'Laurent Visquel?'

'Yes.'

'Why would he destroy your samples?'

'Perhaps you should ask him.'

He made a note. 'And what does our ... missing person have to do with the samples?'

'Whoever was in the warehouse when the explosion took place might have been trying to recover the samples.'

'I see. You mean our victim might be someone who knew you'd taken samples, but didn't know they'd been destroyed?'

'Exactly.'

'Any candidates?'

'Camilla Hatton.' Jaq described the meeting in Café Charlie. 'She claimed to be from Zagrovyl.'

'The company that supplied the ... odd material?'

She nodded. 'Zagrovyl deny that she is an employee.'

'How can I contact her?'

'I wish I knew.' Frank and Cheekbones had kept the card. Jaq copied the details from her phone. 'The number is dead,' she added.

'Pictures? Profile?'

'She doesn't seem to exist online. Not on Facebook or LinkedIn.' Or Hairnet.

'Can you describe her?'

'In her fifties, I guess. Short white hair. Green eyes. Fit. Speaks English with a slight accent, maybe Scandinavian.'

'Is there anyone else?'

'Only if they inspected the return pallet.'

'And if they did?'

She chose her words carefully. 'The driver who collected the reject material, I saw him again when I was in Teesside and—'

Will-O'-the-Wisp interrupted. 'So, he's not the victim.'

No, but he's dangerous. He was working with Mr Beige. With Zagrovyl. She wanted to scream it out, but stopped herself. That route led to a corpse in Lake Coniston. If Will-O'-the-Wisp contacted the Cumbrian police they would demand that she return to England. She couldn't risk that. If the English police weren't going to help her, she needed the Slovenian force on her side. At least while she gathered evidence to support her case.

'Anyone else who might have known?'

She thought back. 'Stefan, the security guard.'

'The one who was injured in the first break-in?'

'Yes.'

He opened the folder and checked his notes. 'Stefan Resnik was still in hospital on the Saturday of the explosion. He is due to be released today. So, he's not our victim. Anyone else?'

Jaq shrugged. 'Someone else connected with Zagrovyl who was worried about what was in those samples?'

'What do you think was in the samples?'

Jaq hesitated. She appraised the man sitting opposite her. Too soon to share her worst fears? Not ammonium nitrate or urea or urea nitrate. Perhaps DMC, N,N-dimethylaminoethyl hydrochloride, or its analogue, DEC. Precursors for drugs used to treat heart disease but, like all Schedule 2 chemicals, a potential chemical weapons precursor as well. Could Will-O'-the-Wisp handle this information? Or might he react the same way the English police had?

A uniformed sergeant entered with a document for him to sign and the moment was lost. Hold your fire. Take your time. Lead them to that bastard, Frank Good.

'Can we come back to that?'

He frowned. 'If you insist.'

'Whoever perished must have had access to keys.' She explained the security protocol and the two sets of keys.

'Could they have used your keys?'

'Why not my boss's keys?'

'Laurent Visquel's keys were in a police safe. I can personally vouch for that.'

'And my keys were with me at all times.'

Will-O'-the-Wisp flicked through the pages of his report, the thin paper crackling. 'As I understand it, these are special keys, difficult to copy. But just for the record, is there anyone who might have had access to your keys long enough to try?'

She hung her head. No point in pretending she hadn't considered the same thing. The reason she had gone to Teesside. Perhaps this young policeman could find Camilla where she had failed. Perhaps Camilla had already been found, what was left of her. She bowed her head.

'Camilla Hatton,' she said.

'Camilla again.' The detective made a note, the pen rasping against the paper of his notebook. 'Well, she should be easy to rule in or out. We'll have the first lab results soon. If the DNA is female we may have our victim.' He looked up. But let's explore all avenues in the meantime. 'Anyone else who could have had access to your keys?'

Jaq thought about it and then shook her head.

'What about your . . .' he looked up from his papers, '. . . boyfriend?'

Memories of the Snow Science kangaroo court twisted at her stomach, the contemptuous words of the chairman. *Shacking up with strangers. You are a grandmother, for goodness' sake.* What was it about women and sex that made men so angry? Doesn't it always take two to tango? She thrust her chin forward and jerked her head up to meet his eyes. Calm. Clear. Candid. Pools of limpid green. Just an officer of the law doing his job. But he was on the wrong track with Karel.

'I'm not judging you, Dr Silver. But what do you know about the man you met on the night of the karaoke party?'

Plenty. 'Karel Žižek.' She gave his work address. 'He's a ski instructor. But he was with me during the break-in and . . .' she paused as the kitchen memories came flooding back, '. . . when the explosion happened.'

'But he could have copied them and given them to an accomplice?'

'As could any one of Laurent's girlfriends,' she said. 'Or a policeman. Anything is possible.'

'Fair enough.' The detective snapped his notebook shut and gave her his card. 'I'll talk to Dr Visquel and the staff sergeant who took charge of his keys, but I'll also need a statement from Karel. Don't tell him about this new development. Just ask him to call me.'

Jaq put the card in her pocket. 'As soon as we finish here,' she stood up, 'I'm going back to Kranjskabel.'

To find out what really happened.

Thursday 17 March, Ljubljana, Slovenia

Boris lit a cigarette. He leant against the stone wall under a Roman arch and watched the citizens of Ljubljana go about their business. From the shadows he could see the gateway to the Specialna Enota Policije.

What was Silver doing in there? What was she telling them? How much did she know? How long did he have?

He'd left England fast. The police had stopped his hire car at the junction where the pass from the valley joined the main road. Asked him if he'd been out walking. Whether he'd witnessed anything in the lake. He acted the dumb tourist. Seen nothing. Heard nothing. They let him go, but it was a close-run thing.

After that he tracked her mobile phone signal. Nothing easier to follow. At least, where there was reception. Not like in godforsaken, waterlogged Cumbria.

It had been a good plan. Silver had ruined it.

So much for a thirty-minute run. Liar. She took more time. Time in which everything unravelled.

The man in the beige raincoat wasn't the problem. Boris spotted him long before he stumbled from the steep path onto the beach, his smooth city shoes slipping on the mossy pebbles. Boris watched the dickhead shout out, pull off his hat, raincoat, jacket and shoes, preparing to jump in and save the boy. With the distraction in the lake, it was easy to run up behind him. *Kokot!* Poor, stupid hero.

Boris felled him with a blow to the back of the head. The man toppled straight into the shallows.

'You mad bastard!' He surfaced, gasping. 'What sort of monster uses children as bait?'

So not just a passer-by.

'Mummy!'

Not much time. Boris pulled the man back onto the beach and hit him again and again, twisting him onto his front, straddling his back, pinning his arms and forcing his mouth and nose into shallow water. Remaining there, panting, long after the man had stopped thrashing. Heart racing with the effort. And the excitement.

Where was Silver? She must have heard the cries by now. He rolled the body into the water, kicking it out towards the depths.

Still panting, he checked the pile of clothes for identification, keeping one eye on the path. How did the man know the child was bait? Who else was watching Silver? He reached into the inner jacket pocket. One driving licence and several cards in the name of William Sharp. A notebook. Detailed surveillance notes. This one had been watching Silver for a few days. He put everything back.

'Ben!'

A shout from above. At last. Silver was coming! Getting closer now. The trap was sprung. He retreated to his hiding place.

It was a great plan. A cold, deep lake. Only one way down to the water and he had it covered. Only one fucking way, unless you were off your fucking head.

How was he to know that the *kurva* would be crazy enough to dive off the sodding cliff? Fucking mental.

He watched in disbelief as she passed over him, momentarily awestruck.

He didn't give up, even then. She had to come back this way. With or without the boy. Judging by the amount of water the boat had taken on, it was going to be without. He'd get her on the way back. Exhausted. Distraught. Easy to overcome. Drown her in the shallows and then send her back to the depths.

A tragic accident.

It was still a good plan.

Until Boris saw the fit bloke in the red kayak powering across the lake. Ben's father. *Suka.* That's when he knew he'd missed his opportunity. And when he heard the siren, he realised he had to get away. Fast.

It was a good plan. Until Silver fucked it up.

This was the last time he'd let her humiliate him. She'd tricked him over the samples, she'd made him waste the last of his Semtex on the booby trap in the vending machine, and forced him to start smoking again after all the pain of giving up.

He took a puff and dropped the cigarette end, grinding it with his boot, squashing and flaying it.

This time, she couldn't escape. This time he wasn't just going to kill her, he was going to punish her first.

And that meant watching and waiting for a little longer.

Boris had a new plan.

Thursday 17 March, Ljubljana, Slovenia

Jaq took the bus from Ljubljana to Kranjskabel. As the bus headed north into the mountains, she opened the white envelope containing the forensics report. Although it made for depressing reading, the results were largely straightforward, unremarkable. Except for one thing.

The report listed the discrepancies, comparing the records she had supplied at the time of the break-in with the inventory taken after the explosion. It showed that all the detonators, fuses, fuel oil, cartridges, primary, secondary and commercial explosives – nitroglycerine, dynamite – were unaffected. The limited inventory and careful storage in segregated areas had worked. Safety by design.

Only the Zagrovyl material had gone.

Explosives 101, one of the many courses Jaq taught to young engineers, defined an explosion as the release of energy so sudden that it causes a shockwave.

Of the three types: atomic explosions – like the runaway reaction in Chernobyl; physical explosions – molten lava from the Krakatoa volcano cascading into the sea, vaporising a cubic mile of seawater and creating a blast wave felt three thousand miles away; and chemical explosions, the latter were the easiest to control.

Artificial chemical explosives rely on rapid oxidation, the reaction between nitrogen and oxygen, for example. The reaction happens in microseconds, the temperature shoots up to thousands of degrees, the pressure soars to hundreds of atmospheres and all that liberated energy needs somewhere to go. Bang!

But pure ammonium nitrate was extremely stable. A propellant, not a primary explosive. When used to shift snow, they had to mix it with fuel oil and other, more energetic, ingredients. Something else had set it off.

Judging by the damage, the detonation had started in the area of the vending machine. An electrical short? Exploding coffee? Impossible. It would only have a fraction of the energy needed to create a secondary explosion.

And this was the curious thing: the police found traces of RDX and PETN. She scratched her head as she checked the detailed analysis. How could they have found decomposition products of Semtex when she never stored Semtex at Snow Science? The plastic explosive, invented in the Czech Republic, beloved of terrorists the world over, was of little value in avalanche control.

One thing was certain: no evidence of negligence. This explosion was not accidental. Someone had set it off deliberately, messed it up and perished in the act.

Ping! Jaq checked her phone. A text from Natalie. Hairnet had come up trumps. Jaq clicked on the link. Camilla Hatton had never attended a salon in Teesside, but there were bookings every six weeks at a salon in Mölndal, Sweden and then more recently Den Haag in the Netherlands. And tomorrow, her heart skipped a beat, a booking in Kranjskabel.

Jaq closed her eyes. If her suspicions were correct, that was one appointment the white-haired woman in turquoise salopettes would never make. So desperate to destroy the Zagrovyl material in the Snow Science warehouse, she had destroyed herself in the act.

The sun slipped below the horizon leaving a pellucid sky, periwinkle blue, above mountains fringed with rose gold. As the bus swept into Kranjskabel, a tall figure with golden ringlets emerged onto the wooden balcony of a ski school. Jaq's heart sank. Since visiting Johan, she had avoided thinking about Karel. The relationship was absurd, the age gap too wide. They had nothing in common except skiing. She knew what she had to do, but the realisation saddened her.

Karel didn't notice the bus, lost in conversation with two men. Not customers. Bearded and heavyset, they wore suits, not the garb of typical ski students. Not policemen, either. Something about the body language was wrong.

The bus drove on, buildings obscuring her view. She was the last passenger to leave the bus at the terminus, and by the time she doubled back, the men were gone.

Karel's face lit up as she approached; the guilt made her feet slow and heavy. After the formal kiss, she stepped aside. He frowned and took her gloved hand. 'I was worried you wouldn't come back.'

A group of teenage girls emerged from the ski school. She couldn't tell him here. It would have to wait until they were somewhere private. She owed him that at least.

'The police need a statement.' She handed him Wilem Y'Ispe's card.

'Has something happened?'

What had Will-O'-the-Wisp said? *Don't tell him about the developments.* In other words, don't tell him about the body. She brushed a strand of hair away from her face. 'They want to know if I let the keys to the explosives store out of my sight. I guess I did when I was . . .' She hesitated, searching for the right words. 'When I was with you. I'm sorry, but they need to hear your version.'

'So best not to mention my sleepwalking habit?'

'Don't joke,' she said. 'This is serious.'

Karel grimaced. 'I'll call him when I finish here.' He nodded at the waiting girls. 'I have a class now. Can I come over later?'

No point delaying the difficult conversation. 'Sure.' She dropped her eyes. 'Call me first.'

Jaq shuffled through slush towards her flat. She pulled out her keys, light and insubstantial now the bulky work set had been removed. An image of Sheila's bent head sprang to mind, her erstwhile friend unable to make eye contact as the formalities of suspension were executed, the Snow Science keys swapped for a paper receipt.

Wait a minute.

Sergei. When he left Snow Science, did he hand in his keys? If he had, it would be recorded in his file. If he hadn't, there might be a third set of keys. And that opened all sorts of other possibilities.

Her hand reached for her phone, to call Detective Y'Ispe. The police could subpoena Snow Science for Sergei's HR files. Or was there a faster way to find out, before Laurent destroyed more evidence? She dumped her suitcase in the flat and sprinted back into the street, reaching the stop just in time for the shuttle bus trundling up the hill.

The Snow Science complex loomed silver and grey against the night sky.

'Good evening, Dr Silver.' Patrice, the relief shift security guard was on duty again. 'I see the police still have the warehouse cordoned off. Any news? I'm just back from holiday.'

Good, he didn't know that she was suspended. Would the security system stop her? Only one way to find out.

'Evening, Patrice,' she said. 'Hey, I left my pass behind. Can you let me in?'

'Sure thing, Dr Silver,' he said. 'I can give you a temporary pass.' He took a blank from a drawer and pushed it into a machine before tapping something on a keyboard.

'You know your employee number?'

Jaq thought fast. Sheila was efficient, she would have blocked the account by now. But employee numbers were sequential. Rita had started on the same day, registered just before her. She subtracted one and gave him that number instead.

'Here you are.' He handed her the pass without checking the name. 'Valid for twenty-four hours. Anything else you need?'

Could she push her luck a little further? Nothing ventured, nothing gained. 'The office keys,' she said.

He unlocked the key cabinet. 'Which office?'

Jaq scanned the rows of keys. The HR information was kept in Laurent's office. She pointed to Key No. 1.

'Sign here,' Patrice pushed a thick book towards her. The last signature was logged at 5 p.m. She flicked back a few pages. The same signature every morning and evening. Rosa, the cleaner. Nothing else on the date of the break-in. Nothing else on the date of the explosion.

Icicles fringed the covered walkway leading to the office block. The heat of the sun melted the snow by day; it dripped from the roof and froze again at night, leaving hanging ice daggers with needle-sharp points that glinted in the brilliant white light as the halogen security floodlights flashed on and off, the motion sensors tracking her progress.

At the end of the walkway, the office block lay cloaked in darkness. Laurent never worked late, and most of the team followed his example. The key slid into the lock of the outer door. Jaq checked behind her; nothing but a distant orange glow from the security cabin. She opened the door and listened; nothing but the tinkling of stalactites above, the distant hoot of a snowy owl behind. At the inner door, her hand hovered over the light switch and then withdrew. No. She knew her way. Best to draw minimal attention to her presence.

The door swung closed behind her. A whorl of warm air swirled down the corridor, and with it a rustling and chattering. Jaq froze. Was something moving up ahead? She waited. Nothing; just the thud-thump of her heartbeat. She took a deep breath and advanced. The outside floodlights clicked off. She swallowed, her mouth suddenly dry, then continued more slowly, one arm outstretched, fingers skimming the wall. Shadows scurried across the ceiling, bounded by an eerie green halo. Something was sliding towards her. Something or someone? Her hand brushed against softness. She lurched to the other side of the corridor. Instead of hard plaster, she stumbled into yielding, smothering shapelessness.

A scream escaped her lips as cool limbs wrapped themselves around her. She flailed against the enveloping form, punching it away. A metal frame clattered to the floor.

In the faint green glow of the emergency exit sign, the reflective strips of twelve snowsuits blinked at her from where they had fallen, scattered across the corridor floor, malevolent cyborgs vanquished.

Feeling foolish, Jaq righted the frame and returned the garments to their hangers. She used the light from her phone to locate Laurent's office, her hand shaking so hard she could barely fit the key in the door. Inside his office, she locked the door behind her and leant against it, wiping cold sweat from her brow, breathing hard. What a twerp. What on earth was she doing here? Breaking and entering. She wasn't cut out for this sort of cloak-and-dagger stuff. The sooner she got out of here the better.

She closed the blinds before putting on the light. Laurent still had his 5S plan pinned up on the wall. A rectangle for his phone. A line for his pen. More shapes for his laptop, monitor, mouse and keyboard. Replicated perfectly on his desk. He was taking this new fad seriously. She sat on the edge of his padded leather chair, wrinkling her nose at the smell of stale cologne. His massive walnut desk had five drawers. All locked. Spare key? She ran her fingers under the desk, under his chair, checked under the rug, above the door, behind the filing cabinet: nothing.

The connecting door to his secretary's office was not locked. Sheila's desk was piled high with papers. Either she hadn't caught on to 5S with the same enthusiasm, or she had more work to do than Laurent. Probably both.

Jaq sat at Sheila's desk. The fabric-covered chair creaked and tilted as she rummaged through the paperwork, careful not to leave anything out of place. A pile of grant applications: cryospheric data project, low-temperature fusion, artificial glaciers, pumpable ice, energy storage, 3D ice printing. An in-tray full of administrative documents: travel requests, train tickets, hotel bookings.

Some correspondence. No personnel files. Nothing remotely confidential. Sheila's single drawer was locked, but the desk was flimsy laminate. It took only a few sharp tugs to force it open. And right at the back, in a brown envelope taped to the far corner of the drawer, were the keys to the secure store.

She returned to Laurent's office and unlocked the store. Inside was a safe and two filing cabinets. The HR files were ordered alphabetically. Jaq didn't bother with her own file; she went straight to the section for ex-employees. There weren't many. The Slovenia branch was relatively new, and people tended to stay with Snow Science. The work was interesting, the pay and conditions good. Laurent might be lazy, but he was not stupid. He let the scientists and engineers get on with their work without too much interference. So long as they brought in grant money, published papers which credited him among the authors and let him attend international conferences in exotic places, he left them alone.

The third file she found, the cardboard scuffed and coarse against her fingers, was labelled Sergei Koval. Who was the mysterious Sergei? Why had he left Snow Science?

Jaq opened the thick folder. Did she have time to take copies? The photocopier in Sheila's room would need an access code to ensure it was charged to the right department. Laurent was a stickler for cost centres. It was safer to take pictures on her phone. She laid the file on Laurent's desk and switched on the anglepoise desk lamp.

The first page had a list of contents: employee registration, bank details, medical certificate, flying certificates, explosives licence, CV, references, insurance, salary letters and disciplinary meeting notes – lots of them. Sergei was trouble. She took pictures of anything interesting as she worked through the file.

As she skimmed through the information, her interest was piqued. Sergei Koval, born 13 February 1955 in Donetsk, Ukraine, had flown with the Soviet air force, studied engineering in Kiev and worked for a series of major international construction companies, moving further west with each job. He gained indefinite leave to stay in the European Union when Slovenia joined in 2004. One page caught her attention. The certificate of a medal of honour for services after the Chernobyl accident.

Chernobyl. 1986. The world's worst nuclear accident. Sergei must have been thirty-one when the nuclear reactor exploded. Was he mobilised to assist with the 'clean-up'? Had he been one of the helicopter pilots charged with dumping tonnes of sand and boron onto the smouldering reactor? If so, he was lucky to have survived. The exposure to radiation was high, and the death toll among the first responders had been terrible.

A noise outside. Jaq froze. Footsteps in the corridor. Pausing. Moving on. Jaq held her breath until she heard a vacuum start up. Just the cleaner in the canteen.

Jaq flicked through the file, taking pictures. Apart from the normal personal data and professional licences, most of the file was taken up with disciplinary investigations. Sergei had a habit of borrowing helicopters when he wasn't on duty.

His experience in Chernobyl certainly hadn't dampened his independence of spirit. Jaq couldn't help grinning as she read the transcripts of the interviews. Not only had Sergei 'borrowed' helicopters at regular intervals, they had found empty fuel cans in the cockpit, suggesting that he had gone on trips that required him to stop and refuel.

He must have been well regarded. Numerous infractions had been ignored before it came to a formal disciplinary process. There was even a written protest from the chief of the air crew about the need for an inquiry at all. The investigation had been blocked at every turn. But eventually Sergei had taken it too far. Absent in a Sikorsky when a major avalanche necessitated a massive rescue operation, he flew in at the last minute. It was only by daredevil flying that he managed to rescue the last of the trapped mountaineers. Someone intervened, ordered a full inquiry. The investigators must have been shocked by what they found in the helicopter logbooks. Trips east. From Slovenia to Serbia, Hungary, Moldova, Ukraine.

What had he been up to? Joyriding? Smuggling? From the transcripts, Sergei wasn't giving anything away. He remained silent throughout. Incredibly, they hadn't suspended him.

But then, he hadn't left a Snow Science party with someone he wasn't married to.

A final disciplinary interview was scheduled, but he never turned up.

Sergei simply vanished.

Jaq reached the end of the file. No letter of resignation. No exit interview. No form to acknowledge receipt of keys. She flicked back and forth until she found the form for key issue and took a note of the serial numbers. Were they the same ones she had been given?

Jaq went back to the filing cabinet and located her own slim file. She was a model employee by comparison to Sergei, at least until the explosion. She found the form receipting back her keys to the explosives store. The serial numbers were not the same as Sergei's.

What about Laurent's keys? Would he have a file on himself? Sure enough, there it was. Some interesting correspondence with a Ladhaki foundation. Dr Camilla Hatton had been telling the truth about that at least. She flicked on until she found the form. His keys had an earlier serial number. She took a photo of the three pages, side by side.

Just as she thought. Sergei had vanished, taking his keys with him. There was a third set of keys out there.

The sensation of weightlessness caught her off guard, her head suddenly spinning. The vice of self-doubt sprang open, the pressure of responsibility released. It didn't have to be her fault that someone had breached security. It certainly wasn't her fault that they died in the explosion.

But she needed more evidence to convince the authorities to

investigate Zagrovyl. And that bastard, Frank Good, who sent men to kill her.

At least one person knew more than they were telling. She found his file and photographed the address.

Time for an honest talk.

Thursday 17 March, Kranjskabel, Slovenia

Jaq barely noticed the stars as she ran down the hill from Snow Science, an urgency in her step driving her on. Next stop, Stefan Resnik, the security guard.

His flat lay on the outskirts of town, behind a shopping centre on the main road between Snow Science and Kranjskabel. She stopped at a supermarket and selected a card with wishes for a speedy recovery. He'd been unimpressed by her last gift of fruit. *Whisky would be more bloody use.* Get the man what he wanted, and maybe he'd tell her what she needed to know. She bought a single malt in a presentation tin.

Jaq cut across the supermarket car park, narrowly missing a van that skidded on the snow near the exit steps. Careless idiots. Stefan's flat was in the basement of a run-down block. The lights were on, but there was no response when she rang the bell. Jaq checked the address and peered into the depths. Maybe he had gone out and left the lights on. Or maybe he wasn't answering.

A van idled at the corner; the engine revved. The same black van that had almost run her over in the car park. Not in such a hurry now. Come to apologise? The windows were tinted, but she could make out three shapes hunched behind the windscreen. She raised her phone and snapped a picture, and another one of the number plate. Moving forward to get a better shot, the van roared into life and sped away.

It was too cold to hang around outside on the street. She pressed each of the bells on the other flats in turn until someone answered.

'I have something for Stefan Resnik,' she said. A blind rattled upwards on the third floor and a lozenge of yellow light spilled on to the street. Jaq held up the bottle in salute, and after a minute the door buzzed open.

Jaq descended into the cold, dark stairwell, using the torch on her phone to inspect each of the four flats in the basement. Stefan's name was on the door of the last one. She placed the bottle behind a utility metering cabinet tucked away out of sight and tugged off a glove to write directions to it on the card. As she bent to slip the envelope under the door, a shadow moved behind the fisheye.

The door flew open. Stefan stood in front of her in a pair of joggers and a string vest. Thinner than before, his face gaunt and unshaven, grey stubble contrasting with the clean white surgical dressing on his head.

His eyes moved from side to side, following her as she retrieved the hidden bottle of whisky. 'Come in, quick.' He grabbed her arm and dragged her into the flat, closing the door behind them.

He ignored her greeting, continuing to peer through the fisheye into the basement corridor. 'Did anyone see you?' he growled. 'Does anyone know you're here?'

'I brought you a card,' she said. 'And a bottle of whisky.'

He locked and chained the door and led her to his living room. A battered sofa stood at right angles to a newer reclining chair which faced a large flat-screen TV across a small patterned rug. A dresser in the corner held a set of mismatched glasses and a strangely incongruous tea set: fluted china cups and saucers in a floral pattern with a gold rim. The table next to the comfy chair was piled high with paperback thrillers, a mug with reading glasses balanced precariously on the top, the metal frames glinting under a floor-mounted reading lamp. The room was heated by a small black stove. The walls were bare, but discoloured squares on the magnolia paintwork hinted at pictures recently removed. A single window high on the wall gave a view of a cinder block wall, well below street level.

Jaq offered him the envelope and bottle.

Stefan threw down the card, grabbed the whisky and took it to the dresser.

'You shouldn't have come,' he said. 'It's not safe.'

She remained standing. 'What do you mean, not safe?'

'There are bad people out there, Dr Silver.'

'Who?'

'I don't know.' He shook his head. 'And I don't want to find out.' He reached into the dresser and drew out two glasses. He poured a couple of fingers of whisky into one and pushed it towards her.

She shook her head. 'No, it's for you.'

He downed the amber liquid in one gulp, sighed and poured another shot. 'I'm not coming back to Snow Science.'

What did he think of her? That her visit was to coerce a sick old man back to work? She wouldn't put it past Laurent to do that, but she wasn't here to do her boss's dirty work. 'You take all the time you need.'

'I'm quitting. Retiring. Going to live with my daughter in the south.' He nodded at a framed picture on the mantelpiece. 'You should quit too,' he added.

'I've been suspended,' she said.

'Because of the explosion?' He tutted. 'It wasn't your fault.' He took another swig. 'Anyone could have broken in. Sergei was careless with his keys.'

'What do you mean?'

'He said they were too bulky and heavy to carry around. When he wasn't in a snowsuit, Sergei only ever wore jeans and a fitted leather jacket, the kind of man who wouldn't be seen dead with a handbag.' He snorted in disgust.

Unlike Laurent. 'So, Sergei returned them to the office safe?'

'Fat lot of use that would be. Since when did the avalanche teams only work office hours?'

'What did he do with them?'

'He kept them hidden under a drain cover.'

Minha nossa Senhora! 'Where?'

'Helicopter landing circle. Outside the warehouse.'

Cabrão! 'Who else knew?'

'Everyone except Dr Visquel.' Stefan shrugged. 'All those stupid rules, totally impractical. Dreamt up by idiots.'

Would anyone believe it? The whole basis of security, bypassed for want of a manbag? 'Will you tell the police what you just told me?'

'I don't care who knows now, I'm not coming back.' Stefan craned his neck towards the window, peering up at the road. 'Did you see that van when you came in?'

Jaq followed his gaze. The van was back, parked in the street, snow tyres and a rusty undercarriage.

'A delivery van, I saw it in the supermarket car park.'

'How many men inside?' The tremble started in his voice, quickly spreading to his thin body.

'Three.'

He groaned; his face was grey, etched with lines of pain. He fumbled with his pocket and pulled out his angina spray. *Pssst. Pssst.*

Jaq guided him to his chair, shocked at how frail he had become in the three weeks since the delivery that had started all this trouble.

He pointed to the bandage on his head. 'Can you get my tablets?'

She found a blister pack of painkillers – paracetamol – and brought them to him together with a glass of water. He took two tablets.

Jaq waited until he'd stopped trembling. 'Do you want me to call a doctor?'

'No,' he said and grimaced. 'Another whisky, please.'

'I'm not sure—'

'I'm sure.'

She poured a finger's worth and laid it on the table beside him. As the colour returned to his cheeks, the shaking stopped. Jaq sat down opposite him. 'Stefan, I need to know what you know. Who broke into the warehouse?' She needed to know a lot more than that. What caused the explosion? Who died? What were Zagrovyl trying to hide? Don't rush him. One step at a time.

'I don't know anything.' He ran a hand over his head, pushing the wispy grey strands away from the dressing.

'Where is Sergei?'

Stefan adjusted his vest. 'Long gone,' he said. 'Back east.'

'Where?'

'Ukraine. A big construction project near Kiev.'

'Has he been back? Could it be Sergei . . .?' She hesitated. Will-O'-the-Wisp Y'Ispe had impressed upon her that news of a body must not be leaked until the police announced it at a press conference. 'Could it be Sergei who broke in? Sergei who set off the explosion by accident?'

'No.' Stefan shook his head. 'Sergei knew what he was doing when it came to explosives.' He finished his whisky. 'A tough guy, but loyal.' He touched the bandage on his head. 'Sergei wouldn't hurt me.'

Jaq knelt on the floor in front of him and took his hand. 'What happened, Stefan? You can tell me. Who broke in? Who attacked you?'

And why? What is all this really about?

'Best if you go.' Stefan squeezed her hand and then disentangled his. 'Thanks for the whisky.'

'If you tell me, maybe I can help.'

He closed his eyes. 'I'm tired. I'm going to take a nap now.'

'But you'll talk to the police?'

He nodded. 'I might need their protection.'

Stefan wouldn't speak to her after that. She tried for a while, but when he began snoring, she called a taxi. She waited until the car was outside the window before letting herself out. The black van was nowhere to be seen.

Thursday 17 March, Kranjskabel, Slovenia

The flat had not changed in Jaq's absence. The same white walls, the same woodchip ceiling, the same Paula Rego ceramic tile depicting a woman leaping over fire, the same books and an orange rope that she'd forgotten to return to Snow Science, neatly coiled on her desk.

But Jaq had changed. She'd blown away the cloud of disgrace that had been hanging over her, banished the black fug of injustice. With this new evidence, the pictures from Sergei's file and Stefan's testimony she had tangible proof of her innocence. Now Will-O'-the-Wisp would have no choice but to investigate Zagrovyl.

She called the detective, but his number was engaged. She left a message for him to call her back.

Her mood shifted. The brief elation was dashed by the prospect of saying goodbye to Karel. She dreaded the approaching discussion, explaining why they had to go their separate ways. How to avoid hurting the beautiful young man who had brightened her life for a while? How to avoid wounding his pride? Perhaps she flattered herself. Had he reached the same conclusion in her absence? Perhaps he was girding his loins, searching for the courage to tell her the same thing. They both knew it was over. A future couldn't be built on great food and mind-blowing sex alone.

While she waited, Jaq flicked through the pictures of Sergei's file to distract herself. What had she missed? She paused at Sergei's life insurance form, zooming in on the contact number for his next of kin. No name, but a telephone number with country code 375. Ukraine? She looked it up on her phone: Belarus, Minsk. Did Sergei's family live in Belarus now? Parents? Partner? Children? Only one way to find out. Her fingers flew over the keys as she entered the number. Then she hesitated. What was she going to say? Just ask for Sergei Koval and see what response she got?

She jumped when the phone rang.

'Hi, have you eaten?' Karel asked.

That flutter in her stomach might be hunger. 'Not yet, but—'

'I'll pick something up. With you in thirty minutes, OK?'

'OK.'

Thirty minutes. Enough time to tidy up, wipe down the kitchen surfaces, dust, run the vacuum cleaner over the carpet, put her clothes away, change the sheets, scrub the bathroom floor, take a shower. Or . . .

Jaq dialled the number on Sergei's life insurance form.

A woman answered immediately. '*Slooshayoo.*'

Belarussian or Russian? The two languages were close.

'*Zdravstvuyte*,' Jaq began.

'I speak English.' The voice was gruff, the accent thick. 'What do you want?'

Jaq swallowed hard. 'I need to speak to Sergei Koval.'

There was a sharp intake of breath at the other end. 'Camilla?'

Result! Play it safe. 'Camilla,' Jaq repeated. Not lying exactly, not claiming to be someone she was not. Just trying to keep the conversation going.

'Thanks God. I begins to worry. Key is here.'

His keys. Damn. If they were in Minsk, how could they have been used for the break-in?

'Where is Sergei?' Jaq asked.

'Sergei is bad boy. It is too much, too dangerous. People ask questions, get suspicious. I will destroy it.'

'No, please don't—'

'Then you come. Bring the money.'

'What money?'

'The money Sergei promised. He calls me his Elena, he ask me to do many things. I do them, but he never pay.'

At last, a name. 'Elena, I'm sorry, but—'

'Come to Minsk, to the nightclub on Ultisa Shishkina.'

A rasp of a match striking and the crackling puff of a cigarette was followed by a long exhale. Elena coughed and then continued. 'Bring a carton of English cigarettes, the gold ones, and a bottle of good foreign vodka. That way, I know it is you.'

So, Elena hadn't met Camilla.

'Ask for the fat Russian whore.' Elena chuckled. 'Come soon, or I destroy it.'

'No, don't do—'

The line went dead.

A Russian prostitute in a Minsk nightclub. So not Sergei's wife, then. Or perhaps it translated badly. Maybe it sounded better in Belarussian. And why did Elena say a key – singular, when Sergei's bunch must contain eight to ten keys – plural? Perhaps it was just her English. Or perhaps Elena was talking about a different key, the thing that Camilla claimed to be searching for.

Or perhaps it was a trap.

One thing was certain: Jaq was not travelling to Belarus to impersonate Camilla and recover any keys. Time to get a grip on things. Camilla had vanished, presumed dead. If Camilla was to be believed, Sergei had also vanished. Only a complete fool would follow in such precarious footsteps. Whatever they were involved with, whether alive or dead, she'd had enough of it. This was too big for her. She was handing everything over to Detective Y'Ispe – Elena's details, Sergei's file with proof that an extra set of keys existed, Stefan's confession about the lax security and her theory that Snow Science was being used by smugglers connected to Zagrovyl to launder intermediates for chemical weapons. This was all too serious for a lone engineer. It warranted a major police operation. A job for OPCW and Interpol.

Karel would be arriving any minute. Best to make their last evening together count.

The moment she turned on the shower, the phone rang. She let it go to voicemail. Steam came in clouds from the shower cubicle.

She stripped and stepped in. Aaah. Nothing quite like the flow of warm water over cold skin. Through the pitter-patter of the shower on the plastic tray, the phone rang again.

Karel was on his way. Despite her best intentions, a quiver started from deep inside her, a wave of desire pulling at her nipples and making the corners of her mouth turn upwards as well. She rinsed her hair and bent for the soap, lathering her skin, enjoying the sensation, the anticipation.

Noises on the staircase outside. When the door flew open, she was half turned away. A rush of cold air, the unfamiliar smell of stale cigarette smoke, made her whirl round and drop the soap. Her hand flew to her mouth. Two men stood in the doorway to her bathroom.

With guns.

Thursday 17 March, Kranjskabel, Slovenia

The tiny bathroom offered Jaq no place to hide. No shower curtain, no cubicle door. A shower head jutted from the wall beside the basin and the water ran into a drain in the corner opposite the toilet. Jaq stood naked, facing two complete strangers. One of them waved a gun at her.

She grabbed a towel from the rail, covering herself as she pressed her back against the tiled wall.

'Get out!' she said.

The men didn't move. One was square and heavyset; he stood with legs apart, chin on chest, the air of a nightclub bouncer. The taller one stroked his short red beard. He spoke in Russian.

'*Priyti bistro.*' You're coming with us.

What could she use to defend herself? They stood between her and the kitchen knives. Perhaps just as well. One semi-naked woman against two armed men. Not such great odds. Was this what had happened to Sergei? A visit from strangers with guns? Followed by . . . what? She shuddered under the towel. 'Who are you?'

'*My zadadim voprosy.*' We'll ask the questions. Redbeard chucked a random assortment of clothes in her direction. '*Odet'sya, potoropit'sya.*' Get dressed. Hurry up.

She dressed slowly under the towel. Pants, jeans, shirt. Was that a noise at the door? Karel must be on his way. All she needed was a distraction. They were bigger, but she was faster. Come on. Play for time. She discarded the first pair of socks and asked for another.

'*Bystryy.*' Bouncer grabbed her arm. He hauled her towards the door.

If she left with them, would she ever be seen again? Would her Snow Science file end like Sergei's? The fear descended – a glacial mist, freezing her blood, locking her joints, hardening her muscles. She fell forward, her bare feet dragging across tiles and onto the carpet as Bouncer barrelled forward. The painful heat from the acrylic fibre against her skin catalysed a rush of adrenaline. She opened her mouth to scream, but Redbeard clamped a hand around her face before she had reached top C. She bit down hard on his hand and stabbed an elbow into his diaphragm.

'*Blyed!*' he yelled as she twisted free and made a dash for the door.

Bouncer launched himself after her. She was out of the door, kicking it closed behind her so he slammed into it. Taking fast, deep breaths, she vaulted over the first balcony and raced down a flight of stairs. Quick. Where to? Head for the back exit. Slam. The flat door flew open and bounced against the wall, but the few seconds had given her a chance to get out of sight.

She moved more slowly now, so they wouldn't hear her. The stairs were wet with slush, melting snow from the boots of the kidnappers. Her bare feet skidded across the steps, the freezing metal edges catching and searing her skin. On the second floor, light pooled under a front door. Loud music from a TV show. Butter and onions. Someone was cooking in front of the TV. Should she cry out? Would they hear her? The men upstairs certainly would. If she battered on that door with her fists and hollered at the top of her voice, would a stranger open it before the men reached her? And if not? She was almost at the back stairwell, the private one that led to the drying room and ski rack. Her best chance of escape was silence, to slip out the back. Her flat was only minutes from the centre. Bright lights. People. Somewhere she could blend in and lose her assailants.

Jaq slipped through the door and into the back stairwell before the men reached the second floor. She could hear them shouting. She jumped the last few steps and landed awkwardly. *Caraças!* She didn't need a twisted ankle right now. She hobbled past the ski rack and tugged at the outside door.

And stopped.

Boris stood outside the back door, his black-bearded face lit by a glowing cigarette, a gun in his other hand.

'Silver,' he said, and snarled. 'You won't be able to dive over me this time.'

Santissima Trindade. She slammed the door and put her back against it, her heart thudding louder than a door knocker. Would he shoot her? In public? In cold blood? Could she run past him? Or get her skis and make a dash for it?

Too late. The other men had discovered her escape route. They banged down the stairs above her. *Merda.* She grabbed a ski from the rack and pointed the sharp end at the corridor, preparing to defend herself. She didn't hear the back door open, but she smelt fresh tobacco smoke in a gust of cold air as the barrel of a gun pressed into her back.

Jaq froze. Redbeard nodded at Boris behind her and removed the ski from her hands. Boris kept the gun to her back while Bouncer shoved a ski sock into her mouth.

Jaq struggled not to choke. The woollen fabric made her retch. She couldn't cry out, but she could conserve her energy. Concentrate. Focus. Redbeard and Bouncer held an arm each and lifted her off the ground, carrying her through the back door into the street. Her toes skimmed the snow as they hauled her round the side of the building to the waiting black van.

She kicked and squirmed, twisting her neck in the direction that Karel would approach from. Where are you when I need you? Was that a shadow lengthening under a street lamp? A tall man with bright curls just round the corner? She heaved and managed to spit out the sock, call out to him. 'Help!'

Boris punched her in the stomach, and she collapsed forward. *Poça!* Breathe. You must breathe. Don't panic. Breathe. Control. Inhale. Expand your chest. Create a vacuum. Let the air in.

The breath came finally in a great rasp of freezing air. But the shadow had gone. Had she imagined it? Or had Karel run away?

No one had rushed to her rescue. No one was going to save her. Boris slapped her hard and stuffed the sock back in her mouth. She couldn't cry out. The wail of a distant siren made the noise she wanted to.

Her cheek stung, and her eyes watered at the smell of diesel from the spluttering black van. A screech of metal on metal as the side door was yanked open. Boris was in the driving seat. Redbeard bent down and grabbed her ankles and Bouncer caught her arms as she fell backwards. As they carried her towards the open van, she looked up to the sky. A rash of stars in a clear black canopy. A full moon. *Meu Deus.* Was this the end? Was this the last thing she would see?

The siren was getting louder. Jaq stopped struggling and let herself go limp. She waited until Redbeard had loosened his hold on her legs to climb inside the van. Then she contorted her body into a backflip, kicking Redbeard under the chin so he fell inside the van, running her bare feet up the open door frame and pushing off the top, using Bouncer as a pivot for her shoulders so that she somersaulted right over him. He slipped and fell to the ground. As he lunged out to grab her, pull her down, she leapt over his head and dodged behind the van. Heart thumping against her chest, she sprinted into the snowy street.

Flashing blue lights sped down the hill. Jaq charged upwards, swinging her arms in broad strokes. Would Boris shoot her from behind? The van engine coughed and revved as it screeched away, down the hill, away from her. She let out a long breath and wilted, her muscles trembling with relief.

But there was a new danger. Headlights. Coming fast. The car was skidding on the ice, a fan of slush, a screech of brakes. Had they seen her? Could they stop in time? Suddenly Karel was there, running towards her, scooping her out of the path of the vehicle. The police car swerved to a halt a few inches away. A uniformed officer jumped out.

They made an odd couple, Jaq barefoot in the snow, wet hair

and jeans, in a crumpled shirt with half the buttons missing, and Karel in full snow gear with his arms wrapped around her.

'Jaq, what the hell are you doing out here?' Karel released her.

The uniformed officer squinted up at the block of flats and then down at the official document in his hand. 'Are you Jaqueline Silver?'

Jaq appraised him. Her relief at the arrival of the police car was tempered by a new danger.

'Yes.'

The uniformed officer stepped forward. 'Jaqueline Silver.' His voice carried loud and clear. 'You are under arrest.'

Thursday 17 March, Kranjskabel, Slovenia

The snow fell softly, covering the handprints where Jaq had tumbled into the road, covering the line made by her toes where she had been dragged to the van, covering the footprints of Boris and the Russians, covering the tracks of the black van itself. The street lamp on the corner cast a matte orange glow over the snow, sucking all the sparkle into its low-pressure sodium chamber, the maximum lumens per watt, in sharp contrast to the twinkling white reflections from the stroboscopic blue light on top of the police car. The siren wailed to silence. Jaq's breathing slowed to normal. Everything peaceful. The snow muted noise, absorbed energy.

Snowflakes rested on the fur collar of the uniformed officer who held out a warrant for her arrest. Big flakes, each one a unique crystal. Diamond dust. Stellar dendrites. Fern-fronded stars.

Jaq stood barefoot in the snow as he read her her rights. Under arrest? For what? She ran a hand through her damp hair and shivered.

'Those men tried to abduct me,' she said.

'What men?' The policeman addressed Karel. 'Did you see anything, sir?'

Karel shook his head. 'I arrived at the same time as you.'

Jaq glanced at him. That shadow under the street lamp. Could he have been lurking there all the time? Too scared to confront men with guns? And now too ashamed to admit his cowardice? Or worse. What could be worse?

Jaq stood her ground. 'Three men. I know one of them. Boris ... I don't know his surname ... he's a delivery driver for Zagrovyl. Two accomplices, Russians. In a black Volkswagen van. They broke into my apartment and tried to kidnap me . . .' Jaq trailed off as she recognised the weary look in the officer's eyes.

He raised an eyebrow. 'Come along, miss. You can tell us all about it at the station.'

She whirled round towards Karel. 'Boris was with the same men you met outside the ski school this afternoon. I'm sure of it.'

Karel shook his head. 'Jaq, I'm sorry. It's best if you go with the police.'

You bastard. Some knight in shining armour! Fit and strong and cowardly. Jaq glared at him. 'Boris is a regular driver. Snow Science will have records.' She appealed to the inspector. 'I need my phone.' She had a picture of their van from outside Stefan's. The number plate. Evidence. She lifted her bare feet, bright red from the cold. 'I need socks, shoes and proper clothes.'

'I'm sorry, miss, you need to come with us to the station.'

'No. You need to secure the flat,' Jaq said. 'It's a crime scene.'

'And what crime would that be, miss?'

'They're trying to kill me!'

'You come along with us, and you can file a report.' He gestured to the police car.

'I'll go and get your things,' Karel said.

'Be quick, sir.'

'My bag, too . . .' Jaq shouted after him.

'This way, miss.' The uniform opened the back door and helped her inside.

When Karel came back down he had her coat, sweater, socks and shoes.

'My phone?'

He shrugged. 'No sign of it.'

Jaq tried to remember the sequence. Her phone had been ringing when she was in the shower. If it wasn't with her in the bathroom, then it must have been in the main room. Had the Russians taken it? Her shoulders drooped. All the evidence. All the information from Sergei's file, her proof that there was another set of keys, the investigation reports and the van number plate. All gone.

'Shall I come with you?' Karel asked.

She shook her head. 'Stay here. There must be fingerprints, forensic evidence.' She turned back to the policeman. 'Where are you taking me?'

'Jesenice Police Station.'

'No.' She shook her head. 'Take me to Ljubljana. I demand to speak to Detective Wilem Y'Ispe.'

Friday 18 March, Ljubljana, Slovenia

The interview room, in the police station at Ljubljana, had mist on the inside of the windows. Little trails of condensation rolled down the glass but never reached the bottom. Warm enough to evaporate the droplets before they pooled on the sill, but Jaq kept her coat buttoned up. The terrible juddering had eased, but cold had seeped into her bones. She kept her fingers wrapped around the mug of hot chocolate, breathing in the hot, sweet steam of temporary refuge.

'Why am I here?'

'No idea.' Detective Y'Ispe flicked a wisp of fair hair from his forehead, his green eyes shining with concern. 'Are you okay? Do you need a doctor?' Will-O'-the-Wisp, the one policeman who listened to her.

'I need some . . .' why should something so natural be so embarrassing? '. . . some tampons or sanitary pads.' Trust her period to arrive two days early.

Will-O'-the-Wisp looked away, a slight flush at his throat. 'The desk sergeant will arrange something.' He made a brief call. 'Tell me what happened.'

The furrows in his brow deepened as he wrote down the sequence of events.

'I'm not hurt,' Jaq said. 'But someone will be if you don't catch those thugs.'

He took down a description of Boris, Redbeard and Bouncer. 'Did anyone else see them?' he asked.

'I think Karel was talking to the two Russians earlier in the day.'

'I'm on the case.'

A female PC accompanied her to the toilets, stern-faced but

with a pack of pads. The old-fashioned kind, without wings or adhesive. Only slightly better than a wad of toilet paper. It would have to do for now.

Back in the interview room, Will-O'-the-Wisp was on his feet.

'Better?'

Not really; her lower back ached, she wanted to wash and then curl up in a warm bed with a hot-water bottle. But it would have to do for now. 'Yes, thanks.' Time to change the subject, move the conversation on. 'I read the forensics report. The explosion was no accident.'

'That was the chief inspector's conclusion as well. He's on his way to explain what's happening.'

'There's another set of keys,' Jaq continued. 'Sergei Koval, my predecessor at Snow Science, he had a third set.'

'How do you know?'

'Get his personnel file from Snow Science. And talk to Stefan Resnik, the security guard.'

'Jaq—'

'And something else,' she interrupted. 'I think Zagrovyl is smuggling banned chemicals through Snow Science.'

'*Jebemti!*' He reeled back. 'That's a pretty serious accusation. Who else knows about this?'

'Please, you need to contact OPCW direct.'

A burly, broad-shouldered man with a square jaw marched in and turned hard black eyes on Jaq. Will-O'-the-Wisp saluted and pulled out a chair for his superior, introducing him as Chief Inspector Goran Trubor.

'Go.' The chief inspector dismissed Will-O'-the-Wisp with a flick of the hand.

'Sir—' he protested.

'I'm taking over. You have other work to do.'

Jaq shivered. A freezing draught blew in as the only policeman she trusted left the room. The door remained open to admit a uniformed officer escorting a man who introduced himself as her lawyer. They must've scraped the bottom of the barrel to find him. Elderly, rumpled suit, stubble on his chin, he stank of cigarettes and yesterday's alcohol. Without the warmth of green-eyed Will-O'-the-Wisp, the temperature in the room plummeted. Chief Inspector Trubor rolled up the sleeves of his striped shirt and barked into a tape recorder the date and time and names of those assembled.

'We are investigating the manslaughter of an unidentified person, found in the Snow Science store after the explosion on . . .'

Wait. The professional negligence accusation had morphed into a manslaughter charge. So, they knew the explosion was not an accident. Did they still think she was responsible?

'. . . and the murder of security guard, Stefan Resnik.'

Jaq dropped the mug. 'Stefan? Murdered?' Hot chocolate spilled and ran across the table, dripping onto her jeans. *Ó meu Deus, que horror!* 'What happened?'

'We were hoping you might tell us.'

The black van. The same one outside his house. *Credo.* The Russians.

'When did you last see Stefan Resnik?'

'Today.' Jaq glanced at the clock; it was after midnight. 'Yesterday. Earlier tonight. I went to his home.'

'Why?' The chief inspector stroked his chin.

Rumple Stubble held up a hand. 'Could I have a moment alone with my client?'

'It's okay.' Jaq waved him away. 'I only wanted to ask Stefan some questions.'

Rumple Stubble threw his hands into the air.

'And did he answer them to your satisfaction?'

'No. He didn't remember anything.'

'In other words, he wouldn't help you?' The chief inspector smiled.

Oh, Christ, where was this leading? 'I just—'

'What time did you go to visit?'

'About seven o'clock.'

'Did you take flowers, chocolates, perhaps?'

Jaq stared down at her jeans. 'A card and a bottle of whisky.'

'So, you went to see an injured man after he was released from hospital.' The chief inspector opened his palms and turned them to the ceiling, his expression one of incredulity. 'With a bottle of whisky?'

'I wanted to check that he was okay,' Jaq said.

'And to ask him some questions,' the chief inspector reminded her.

She glanced over at Rumple Stubble, but he had his phone out, his fingers flying across the keyboard. Had he given up on her? She swallowed hard. Honesty was always the best policy. 'I wanted to know if he remembered anything about the break-in.'

'I see.' The chief inspector straightened his tie. 'And did he?'

'No.' Jaq sighed. 'He was confused.' Confused or scared?

'What time did you leave?'

'I don't remember. Before eight o'clock.'

The chief inspector drummed his fingers on the table. 'So how long were you there? An hour, would you say?'

'Less, he was tired.'

'I see. So, let me get this straight. You disturb an elderly man on the day he is released from hospital after a brutal attack at your facility.' He stressed the word *your*. 'You visit late at night—'

Jaq protested. 'Seven o'clock is hardly late—'

The chief inspector ignored her and continued. 'You ask him questions. His answers are unsatisfactory—'

'I didn't say that—'

'And a short time later he's dead.'

'Poor Stefan.' Jaq bowed her head. 'How did he die?'

'A massive heart attack.'

Rumple Stubble snapped to attention. 'My client can hardly be held responsible for—'

'Let's wait for forensics to finish,' the chief inspector boomed, big and confident. 'Did he ingest anything else while you were there?'

'Yes. I gave him his painkillers.'

'Enough!' Rumple Stubble stood up. 'You must have had a reason to get an arrest warrant.'

'We had a tip-off.'

'Then we have the right to know the details of the accusation. Who sent you to my client's address?'

The policemen exchanged glances. The chief inspector nodded, and the uniform leafed back a few pages until he found the notes he was looking for.

'We were contacted at 20.09,' he said. 'A member of the public told us that Stefan Resnik had been murdered. Our informant claimed the culprit was about to flee and gave us the address of Jaqueline Silver.' He flourished a handkerchief and dabbed his nose.

Rumple Stubble banged his fist on the table. 'Who denounced my client?'

Another exchange of glances.

The chief inspector rose to his feet. 'That information is confidential.'

Monday 28 March, Ig, Slovenia

The women's prison lay a few kilometres south of Ljubljana. The converted Palace of Ig rose several storeys above an arcade courtyard, surrounded by parkland and forest, contained within stone walls.

Jaq perched on a narrow bed in the spartan remand cell waiting for the police escort. The nights passed slowly, but the days were worse.

Every weekday morning, she was taken in a prison van to Ljubljana police station. Her prospect of release deteriorated with each interview. At first Chief Inspector Goran Trubor only wanted to talk about the explosion, refusing to accept that there were other sets of keys, accusing her of booby-trapping the vending machine to deliberately kill someone. Demanding that she reveal the name of her victim.

Next, he focused on Stefan, producing new testimony from the nurse who asked Jaq to leave Jesenice Hospital, as if that proved Jaq had been trying to murder the security guard for some time.

What would she be accused of today? And what had happened to Will-O'-the Wisp? She hadn't seen him for days.

Wearily, she got to her feet as the police escort arrived, hands outstretched ready for the handcuffs.

The journey from Ig to Ljubljana was brief and uncomfortable in shackles. But it wasn't the physical discomfort she minded as much as the confinement. Through tinted windows she could see the fertile valley blooming as the days warmed and lengthened. She longed to be outside.

Inside the police station, the restriction was worse. Through the small window, too high to afford any view, came the noises and smells of a busy little city. Car horns and market cries, roasting meat and garlic, music and chatter, fresh bread and hops, laughter and spring flowers.

She took her seat in the interview room and waited to have the handcuffs removed. Rumple Stubble sat slumped, half-asleep in the corner, as they waited for Chief Inspector Trubor.

He arrived in jovial mood with two uniformed officers. Experience warned her that his uncharacteristic good humour didn't bode well.

After the usual formalities, the chief inspector resumed his questions. 'Do you suffer from migraines, Dr Silver?'

'No.'

'Do you recognise this bottle?'

Jaq squinted at the evidence bag as he placed it on the table. Inside the plastic nestled a brown glass vial. One she had never seen before. 'No.'

'This bottle of pills was found in your bathroom cabinet.'

Jaq shook her head. 'That's not possible. They're not mine.'

'Wait, this is all most irregular.' Rumple Stubble woke up. 'You cannot search my client's flat without permission.'

The chief inspector smiled. 'She gave us permission.'

The lawyer turned to Jaq. 'Did you?'

You need to secure the flat. It's a crime scene.

She bowed her head. 'Yes.' The word came out as a whisper.

The chief inspector handed the evidence bag to the uniform. 'Sergeant, can you confirm that this was found in the bathroom of Miss Jaqueline Silver while she was in custody?'

'I can confirm.'

'And how many capsules are missing?'

'Eight.'

Jaq shivered. 'What's in the bottle?'

'Ergomar. Each capsule contains two milligrams of ergotamine.'

Ergotamine. The drug synthesised from ergot. A fungus found on rye wheat. A dangerous natural poison. Responsible for the deaths of thousands in the Middle Ages. They called it St Anthony's Fire, an illness that caused hallucinations, a sensation of scalding fire, gangrene of the hands and feet. Sufferers went berserk, writhed in agony, vomited and ran crazily in the streets, pulling off their clothes to reduce the terrible burning sensations in their limbs.

She'd written a paper on it once. A rebuttal of the view that synthetic chemicals were always bad and natural substances were always safer. *Natural Danger.* Jaq put her head in her hands.

'In your interview, you stated that . . .' the chief inspector rustled through some sheets of paper until he found the right point, '. . . you gave Stefan Resnik some medicine on the night he died. Is this the medicine you gave him?'

Jaq shook her head. 'Stefan took paracetamol. And those pills are not mine,' Jaq said.

The chief inspector glared at her. 'The autopsy report came back this morning. A heart attack. Our expert confirmed that this could have been triggered by a massive dose of ergotamine. Four capsules would do it. Eight are missing. We are checking for traces of ergotamine in his blood.'

Jaq put her head in her hands. Poor Stefan. He was already ill. He'd been waiting for a heart bypass. Attacked and hospitalised, it wouldn't have taken much more to kill him. Someone was trying to frame her. Someone who knew what they were doing. But why ergotamine? And how did the bottle of anti-migraine pills get into her bathroom cabinet?

The police? Could they have planted the evidence?

'I want to talk to Detective Wilem Y'Ispe,' she said.

She'd asked for him every day, but this time the chief inspector queried his sergeant.

'Is he back from England?'

England? Her heart sank.

The sergeant nodded and left the room to fetch him.

Will-O'-the-Wisp looked even thinner than before, his face pinched, shadows around his eyes and under his cheekbones. He remained in the open doorway, as far away from her as he could get and still be in the same room.

Her stomach twisted as she scrutinised his face. He held her gaze. A new, hard edge; something had changed.

'Dr Silver.' There was no warmth in his greeting.

'Detective Y'Ispe.' She cocked her head in enquiry. 'Did you get Sergei Koval's file?'

'Snow Science refused our access request,' he replied. 'We'll obtain a warrant in due course.'

'And OPCW?'

'The Organisation for the Prohibition of Chemical Weapons were already aware of your allegations against Snow Science and Zagrovyl.' Will-O'-the-Wisp closed the door behind him. 'They've referred the case back to the national authorities: Slovenian Ministry of Health, UK Department of Energy.'

Kicked it into touch, in other words.

Will-O'-the-Wisp took a step forward. 'The English police wish to speak to you about the death of a Mr William Sharp,' he continued. 'It would have been helpful if you had mentioned that earlier.'

So that's what she saw in those green eyes. Disappointment. Hurt. Suspicion. She looked away.

It came as no surprise when Chief Inspector Trubor informed Rumple Stubble that he would oppose bail. A man had died in suspicious circumstances in England. Two men were dead in Slovenia: the unidentified body found in the ruins of the explosives store and the security guard. Jaq was the prime suspect for all three deaths. She had been the last person to see Stefan before the break-in and the last person to see him alive. The drug that probably killed him had been found in her flat. Poison, the female murder weapon of choice. Explosives, her area of expertise.

Slam dunk.

Friday 1 April, Kranjskabel, Slovenia

Jaq lay beside Karel, listening to his breathing. Every time she moved, he tightened his hold on her. Bodyguard or jailer? Protector or captor? Jaq was no longer sure.

For once, Rumple Stubble had earned his keep: found the flaws in the paperwork. Insisted proper procedure be followed, demanded that Jaq be allowed twenty-four hours to put her affairs in order before returning to Ig Prison to await trial.

Not bail, but temporary release under supervision. They took her passport and sent for Karel, who offered to supervise and promised to bring her in for the formal arraignment.

One last night of freedom. Karel had surpassed himself in the kitchen, but Jaq could barely manage a mouthful. He remained quiet, solicitous, attentive, strangely incurious. Jaq didn't lie to him; told him that she was going away for a long time, that their relationship was over and he should move on. He sought to reassure her, suggested they take it one day at a time. She didn't persist; this was not the time to fight, she had enough enemies as it was. If Karel wanted to pretend to be on her side for a little longer, then she would go along with the charade – whatever got them through the night. For the first time, they didn't make love.

Karel sighed and twitched in his sleep. What was he dreaming about? Every time she moved away, he curled around her, his warm chest against her back, his arms round her waist, his knees in the hollows of her knees. Perfectly tessellated. She could hardly bear his touch.

There were too many coincidences.

The first night, the night at the karaoke bar, was Karel sent to detain her? Did someone want to make sure she didn't meet the first delivery? Keep her occupied until the extra eighteen tonnes were removed?

Who was he meeting in Café Charlie? Camilla? Was it a coincidence he came to Jaq's flat on the night of the break-in? Was Karel's offer to cook her dinner a ruse to make sure she was not around at the warehouse when someone went looking for the samples? And then, after the police cordon was removed, the morning of the second break-in, the day of the explosion, was Karel sent there again to keep her away?

Did he call the police to frame her for Stefan's murder? Did he tell the Russians how to get into her flat? Did he hang back outside, knowing that men with guns were abducting her? And when the kidnap attempt failed, did he plant the evidence that was about to send her to jail?

How could she have been so stupid? How could she not have seen him for what he was – a ski instructor in winter, a mountain guide in summer, an occasional chef, a man for hire. Certainly a great actor, she would never have guessed he was anything but genuine. But how much thought had she given it? When the sex was that good? Face it: she lost all reason when the testosterone mingled with the pheromones.

If someone had put together an identikit perfect man, a honeytrap for Jaq, then they couldn't have done better than Karel: musical, athletic, handsome, gentle, a linguist and a great cook. Had he been hired by Zagrovyl?

She waited until his breathing was regular and slow. Jaq slipped out of bed, placing a pillow in the depression left by her body. She breathed in his scent for the last time: liquorice and lemon – the sweet tinged with the sour.

There were things that had to be done, and she was running out of time.

She dressed quickly, hopping from one foot to the other, attempting to keep the chill from the floor seeping up through the soles of her feet. Karel murmured her name in his sleep but she didn't reply. What was there to say? Once trust is lost, can anything remain?

She peeked through the gap in the curtains at the top of the parked car in the street outside. The police guard was light, two men taking turns guarding the front entrance from the car and the back entrance on foot. They swapped every hour, on the hour. She'd watched them yesterday while she packed: a decoy suitcase with the clothes and books allowed in prison, crates and boxes of possessions to go into storage, and – while Karel was busy in the kitchen – her bag stuffed with the things she really needed.

Jaq checked her watch. At ten minutes before the hour, she opened the front window. Even if he had been paying attention, the man in the car was too close to her building to see the roof.

She converted her bag to a backpack and climbed out into darkness. Thick cloud hid the full moon. Hands scrabbled over roof tiles, uneven and slippy with frost, cold against her cheek as she spread her bodyweight and moved crabwise to the ridge. A light breeze brought a hint of woodsmoke. Jaq peered over. Back Door Man stomped his feet and checked his watch. She extracted the orange rope from the backpack, looped it round the chimney stack and threaded it through the karabiner hook on her belt. Her fingers shook as she tied the double overhand knot. Wait. Listen. The only sound was her thumping heart.

Before the church clock struck five, Back Door Man began to move. It took her fifteen seconds to abseil from the roof and drop into the back alley, another five seconds to recover and stash the rope and then she was off, moving silently away from her flat, sprinting towards freedom.

Dawn was breaking as she boarded the bus. She closed her eyes against the beauty of the mountains. This chapter was over.

Time to start afresh.

Time to take control.

INTERMEZZO

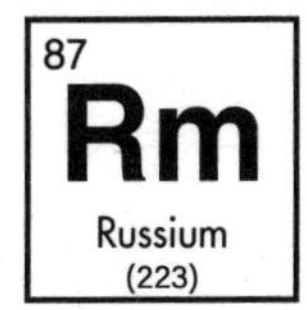

Friday 8 April, Teesside, England

Rain lashed the streets of Middlesbrough. Frank cracked his knuckles as PK parked the limo, extracted the Chariot Cars umbrella and accompanied him to the town hall entrance.

Why was there no news from The Spider? He'd been dragging his bony heels for long enough. It was about time the glorified accountant came up with something worthwhile. The Russian was getting paid ten times as much as Raquel and, so far, had achieved fuck all: unacceptable. Frank dialled and switched the phone to his good ear. 'News?'

'You were right,' Pauk said.

'I am always right.'

'I've tracked down the missing items.' Pauk sounded pleased with himself; he almost chortled. 'I know where they were sent.'

Some progress at last. Frank gazed out at the rain. 'Can you get them back?'

'Best not.'

'What the hell do you mean?'

'This might open up an interesting new . . . business opportunity.'

Frank bit back the angry retort. Pauk was no fool, but the Russian version of an interesting business opportunity was unlikely to delight his boss in the USA, a slave to the Zagrovyl corporate responsibility team. Business prevention team, more like; fucking committees, populated with has-beens and lily-livered lickspittle who would never amount to anything. The only way to get things done was to pounce first and manage the damage limitation after the target had been secured.

'Stop talking in riddles. Explain yourself.'

'Meet me.' The voice was low and rasping.

Frank loosened his tie and unbuttoned the top of his shirt. Perhaps Pauk had more ballerina cousins who needed seeing to. 'Moscow?'

'No. Kiev.' That disconcerting hint of laughter again, part giggle, part smirk. 'I'll take you on a little helicopter trip. To show you something that has to be seen to be believed.'

Frank grunted. 'I am a busy man.' A bell rang; the concert was about to start. 'This had better be good, Pauk.'

'Oh, it's good.' A low chuckle at the end of the line. 'Believe me when I say you will not be disappointed.'

Friday 29 April, Lisbon, Portugal

Jaq sat in the shade of a jacaranda tree opposite the grave of the fifteen-minute baby, the roar of Lisbon traffic hushed by the canopy of mature trees and sprinkled with birdsong. The British Cemetery provided a perfect refuge from the bustling city. Crowded with the dead, empty of the living, lush with creepers and flowering shrubs, birds and insects. Secluded, a place of hibernation, a haven for her thoughts.

The journey from Slovenia had been long and arduous. She took a roundabout route, sleeping on overnight buses and trains. Thanks to borderless Europe, her lack of passport presented no problem so long as she stayed within the Schengen area. All she needed was the old French driving licence in her married name. No one was looking for Jaqueline Coutant. Not yet.

She couldn't go north – to England, so she came south – to Portugal. It was a risk, she knew that. Anyone searching for her would look for family connections. But the police had her passport, and the fastest way to get a new one meant taking this chance.

And when in Lisbon, she had the perfect place to hide. Jaq had started coming to the British Cemetery as a teenager, often staying the night in the caretaker's cottage, fleeing her mother's rages or the disapproval of the nuns, using the underground passage that connected the British Cemetery to the mortuary of the British Hospital, right next door to the convent where Jaq was once incarcerated and where her mother now lived.

If lived was the right word. Her mother no longer moved. She lay ramrod-stiff in the narrow bed, her shrunken body disappearing into the mattress. Once a day she was hoisted into a wing-backed chair next to a shuttered window that opened on to a wall. She barely ate or drank; didn't read or listen to music and never left the room.

Jaq was prepared for the anger, but not for the surge of pity. If nothing else, her mother had always been lovely to look at. It was hard to find any trace of the once-radiant beauty in the withered husk who sat before her now. The 20-watt bulb under a heavy lampshade cast a sickly glow over her sharp bones and translucent skin: an empty shell.

Maria dos Anjos de Ribeiro da Silva, Angie, was born in the highlands of Angola in the province of Bié, to a wealthy Anglo-Portuguese family of colonists who owned a string of successful coffee and rubber plantations.

Her parents expected their only child to make a good marriage inside their narrow community of landowners, a son-in-law to continue the flourishing business.

Instead Angie ran away with a cultural attaché from the Russian embassy. It was an impossible match. Jaq's father was a man of the world: a political intelligence officer – hard, practical and resourceful, a Soviet spy. Angie was a delicate caged bird: beautiful, innocent, impractical – a fantasist raised on fairy stories with handsome princes and happy endings.

Never the strongest, Angie became increasingly unwell as Angola began to unravel. The death of her son, the separation from her husband and the move to Lisbon pushed her sanity over the abyss. But this new form of dementia, if that is what it was, made Jaq feel as if she was being abandoned all over again.

'Hello, Angie,' Jaq said. Her mother had always insisted on first names, detesting *mãe* or Mummy.

No response.

'Your daughter has come to say goodbye.' Sister Magdalena spoke in an exaggerated, loud voice, as if the problem lay in Angie's ears rather than her brain.

No response.

Jaq knelt and extended a hand towards the spotted claw that lay in her mother's lap.

Angie retracted her hand and turned her face to the wall.

'How is she?' Jaq asked as she backed away.

'No change,' said the nun. 'The Lord watches over her.'

Fat lot of use that was.

'Is there anything I can do?' Jaq asked.

'Just pray for her, my child, as we pray for you.'

Good luck with that. Jaq bit back a sharp retort. Her memories of this place were more painful than the young nun could imagine. This is where life began. And ended. Jaq had what she came for: cash from her mother's strongbox, the bottomless fund in a secret box she was never allowed to see; a copy of her birth certificate in the name of Maria Ines Jaqueline da Silva; and a Portuguese passport application in the same name, now endorsed by the priest.

'Bye, Angie,' Jaq said.

There was no response until the door closed. Then came the sobs and the terrible cries. So, her mother had not forgotten. Through her catatonic dementia, there was still no room for forgiveness.

The scent of orange blossom and the sweet song of a thrush brought Jaq back to the present. And the journey ahead. The passport had been fast-tracked by an expert navigator in the choppy waters of Portuguese bureaucracy. Stamped with the visas she would need in her search for Sergei.

She'd lain low long enough. There was nothing to keep her here.

Next week she was flying to Belarus. To a brothel in Minsk. With cigarettes and vodka.

There was a first time for everything.

PART IV: FUGUE UKRAINE

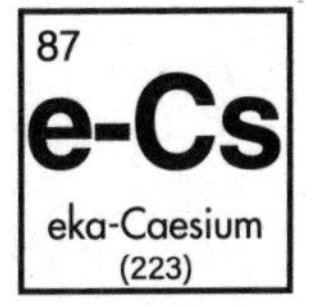

Friday 29 April, somewhere near the Belarus–Ukraine border

The sun sank towards the horizon as Boris approached the complex, a great sphere of red casting its rosy light over the wetlands. A little horse neighed and tossed its mane as Boris passed. The foal continued drinking from the lake, almost hidden in the pale green fronds of willow trees. A beautiful place. Beautiful and deadly.

He'd answered the summons. There was little point hiding from SLYV. Others had tried; it always ended badly.

He lit a cigarette and drew in deeply, savouring the taste, the nicotine hit.

The Spider had agreed to meet with him, to hear his new plan. A good sign; he'd be dead by now if it was up to Mario – the Venezuelan oaf. But The Spider was smart. He'd be impressed. This plan was genius; The Spider couldn't fail to see that.

The grinding of metal on metal made his hair stand on end. Time to go. He nipped the cigarette halfway down, put it in his pocket for later and waited for admittance.

'Boris!' The Spider clapped him on the back. 'You came.'

Boris stumbled in the direction of the seat indicated. The office was simple and functional: desk, chairs, drinks cabinet. The Spider opened the glass doors and pulled out a bottle.

'Vodka?'

You didn't say no to The Spider.

'I've been looking forward to this little chat. A . . . debriefing.' He chuckled. 'I'm keen to understand what happened.' He filled two glasses, pushed one over to Boris and then offered him a cigarette. A black Sobranie. 'Mario tells me you found evidence of a plot against us. Why don't you start from the beginning?'

Bože všemohoucí, he was lucky to have a boss like Pauk. Someone smart. Someone who understood the difficulties of fieldwork.

He lit The Spider's cigarette and then his own.

'*Za vas!*' Pauk raised his glass.

Ukrainian vodka tasted of nothing, but Boris needed the courage. He drained the glass in one gulp and the words began to flow.

It all started with Silver. OK, so Yuri fucked up, but that was easily fixed. If it hadn't been for Silver. Because when Boris went to clean up, he'd found evidence of a tracker.

He waited for respect in the eyes of The Spider as he explained how he'd broken in. He'd watched Sergei unlock the explosives store often enough to know the sequence, and, after some expert persuasion, the security guard revealed where Sergei hid his keys. After a thorough search he'd found something much worse than samples. He had removed the memory stick and key, put the keys back in Sergei's hiding place and left a little surprise for Silver. He'd even left the doors and cages wide open to ensure she got the message.

He tried to gloss over the Semtex trap that had snared the wrong person. Boom! Or his second attempt to kill Silver, in the English Lake. Splash! Or his third attempt in Kranjskabel.

But The Spider wouldn't let him get to the end, to the good part, to the new plan. He seemed fixated on what had gone before, kept asking why. It wasn't going well; Boris wasn't explaining himself properly. The words he'd rehearsed kept coming out wrong. Why, why, why, why, why, why, why.

Everything came back to Silver. It was all her fault, which is why the new plan was so satisfying. He just needed to find the right words to explain it.

'Enough.' With a groan, The Spider twisted in his chair. He bent and unlocked a desk drawer, pulling out a silver flask.

'Boss, I—' Boris's voice cracked, his mouth suddenly dry.

The Spider opened the flask and poured a generous measure into his glass. He pushed the silver flask towards Boris and nodded at the empty tumbler. Boris took a furtive sniff as he filled his glass. Ethanol. Russian vodka? It had to be better than the Ukrainian muck.

The Spider picked up his glass. '*Za vas!*'

Boris took a sip. Yes, this was definitely better stuff.

Anger contorted The Spider's face. 'You didn't clean up at Snow Science, did you?'

Boris took another sip and tried to swallow.

'Work . . . in . . . progress.' Why the fat lips? Why the dry throat?

'And the tracker is still out there, recording our activities.'

'I know . . . how . . . to find . . .' Why the blurred vision? Why the trembling limbs?

'And you let an engineer from Snow Science go to the police? In England. And in Slovenia. Jeopardise the whole operation?'

Why the racing heartbeat? Why the sudden nausea? Why the dizziness?

'Silver . . . part of the plan. You see—'

'I make the plans, not you.' The Spider poured the contents of his own glass back into the flask, untouched. 'How do you like my new cocktail? I took inspiration from your Agent Number Fifteen.'

Quinuclidinyl benzilate. Boris slumped in his seat. No!

'We changed some things to avoid detection. You're a chemist, aren't you? I'm sure you'll appreciate the power of substitution. Personally, I'll stick to vodka.'

The room was spinning. Boris tried to steady himself but his hands no longer obeyed him, his feet were slipping.

Falling, falling, falling.

Friday 6 May, Minsk, Belarus

Poor Minsk, the end of the civilised world, the unluckiest place on the planet. Capital of Belarus – a country ruled by the capricious dictator, Lukashenko – the city had few redeeming features. How could it? Torn apart in the First World War, a quarter of its population killed in the Second World War, the remaining intelligentsia exterminated by Stalin after liberation and then, just when the country was getting its act together, the Chernobyl nuclear reactor exploded. Sixty per cent of the radioactive fallout rained down on Belarus; a quarter of its agricultural land still contaminated.

Jaq took a tram to Minsk Zoo and then walked the last few blocks. And blocks they were – brutalist architecture badly executed. Never had concrete looked so unevenly shabby, so randomly grey. It didn't even have the virtue of uniformity: concrete of the porridge-and-cardboard variety, crumbling away to reveal the rusting metal rebar.

Ulitsa Shishkina stretched out over several blocks in the Syerabranka district, but the nightclub was hard to miss. A giant red neon arrow pointed to a sign in English announcing the Shiskina Model Agency, a low building on the corner with Slanvyy Prospyect – Glorious Prospect. Not so glorious, if truth be told. The club facade was painted red; the front doors were polished steel. And firmly closed.

Jaq slipped in by the back entrance. The fire escape was wedged open with crates of empty bottles: vodka, Crimean champagne, beer. The sunlight streaming through the open door highlighted the imperfections in both decoration and cleanliness. The nightclub had not been designed with daylight in mind.

The door at the end of the corridor opened on to a saloon bar: two threadbare red velvet sofas with fake fur throws and four scuffed leather armchairs around a low table; six high stools beside a polished wooden counter. The sunlight bounced back from the mirrors of the bar, lighting up row upon row of coloured bottles lining the bar shelves like jars of sweets.

In a small alcove, a woman sat at a table in a red velour dressing gown poring over some figures on a clipboard. A large woman, pretty in a soft way. The room stank of stale cigarette smoke. Jaq coughed and held up the duty-free bag with the gold carton of cigarettes and blue vodka box. 'Elena?'

The woman looked up and scowled. 'At last,' she said. 'I thought you were never coming.' She patted the chair beside her and inspected the vodka. Then she turned her attention to Jaq.

'You are Camilla?'

Jaq swallowed. Could she do this? No. Fooling someone by omission was one thing, but she couldn't tell a direct lie. 'My name is Jaq. Camilla asked for my help. We both want to find Sergei.' This was true: Camilla had asked, ordered her, even if Jaq had refused.

Elena wrinkled her forehead. 'But it was you I speak to on the phone?'

'Yes.'

Elena pursed her lips. The deep red lipstick matched her dressing gown. 'How do I know you tell truth now?'

Good question. Diversion required. Jaq pulled out the temporary Snow Science pass Patrice had given her and switched to Russian.

'I worked here, in Kranjskabel.' She pointed to the Snow Science logo. 'Same place Sergei worked.'

Elena narrowed her eyes. 'Were you lovers?'

Jaq crossed her arms. 'No, I never met Sergei. I took over after he left.'

Elena shook her head. 'Then why are you looking for him?'

Why indeed? Keep it simple. 'Something went wrong with the

explosives. I was blamed. But it wasn't my fault, and I need to set the record straight.'

Elena laughed. 'By accusing Sergei?'

Jaq shook her head. 'No. Sergei knew what he was doing. He'd discovered something. Something important. Then he disappeared.'

Elena approached the bar, enormous buttocks and breasts describing their own swaying dance as she moved. She locked the foreign vodka and cigarettes away and returned to the table with a bottle of national spirit and two shot glasses, filling them and pushing one towards Jaq. Breakfast. It was the last thing Jaq wanted right now, but this was a contract. She wasn't going to get any information unless Elena trusted her.

'*Zdarovye!* Bottoms up! Elena finished her glass in a gulp and poured another measure before continuing in Russian.

'Sergei is the luckiest man I know. He should have been dead many times over.'

Jaq kept silent and waited for Elena to continue.

'You know he flew over Chernobyl?'

Jaq nodded.

'He should have died from the radiation dose he received there. Most of the young pilots lasted a few months or years at most. They suffered slowly, horribly. Somehow, he thrives. He has never taken a day off sick in his life. He lives each day as if it is his last. He has no money, always gambling it away. He smokes like a chimney, drinks like a fish, eats like a pig.'

'He's married?'

'Divorced, I think, but he has a new girl every week. He still comes back to me from time to time. I never can resist the bastard.' She reached for the bottle. Jaq put a hand over her own glass.

'Children?'

'Never. He saw what happened to the families of the other pilots, how the sickness spread, how the children suffered.'

'How would I recognise Sergei?'

'A handsome bastard.' Elena stroked the table with the back of her hand, brushing away invisible crumbs. 'Medium height, muscly, pilot's swagger. Plenty of hair, head and body, a bear of a man.'

Elena fanned her face with a beer mat.

'A genius in the air. He can handle any weather. If someone is in a tight spot, they call Sergei.'

'A brave man,' Jaq said.

'No, not brave.' Elena shook her head.

'Why do you say that?'

'Bravery is overcoming fear. Sergei has no fear.' Elena swigged her vodka and sighed. 'Sometimes I think he wants to die. He didn't expect to live so long. Now he has nothing to lose, lives life for the moment with no thought for tomorrow.'

The more she learned about Sergei Koval, the more she understood why Snow Science had hired her. A nice, quiet, married academic – no secrets. If only they knew. 'Where is he now?' Jaq asked.

'No one knows. You're not the first person to come looking for him.'

'Who else?'

'Russians.' Elena spat. 'Bad men.'

The kidnappers? 'What did you tell them?'

'The same thing I'm telling you. I don't know where Sergei is.'

'Why did he leave Snow Science?'

Elena pursed her lips. 'How should I know?'

'Elena, you said he left something for Camilla.'

'Sergei always had a backup plan.' Elena waddled to the cash register, a modern electronic box with a black dongle plugged into the side that glinted with a lilac light. The authorities must record all transactions. Even a brothel had to pay its taxes. Elena reached behind it and retrieved a silver key on a yellow ribbon. She dangled it in front of Jaq.

So not his warehouse keys. This was one Jaq didn't recognise.

'First you give me the money he owes me.'

Jaq shook her head. 'I can't pay.' She held out her hand. 'But if you give me the key, I will try to find Sergei for you.' And for me, she thought, as I've no idea what to do without him.

Elena grasped Jaq's hand. 'If you did know, would you use this key to unlock the secret?'

'Absolutely.'

'Whatever the risk? Despite personal danger?'

Jaq sighed. 'I'm looking at a long prison sentence if I can't find the truth.'

Elena released her hand. 'Then you are the right person.' She turned the key over to reveal a number engraved on the side. 'This is the key to Sergei's secret locker.'

'His secret locker? Where?'

'When you find Sergei, tell him Elena is looking for him.' She hissed something under her breath; her plans for him did not sound particularly loving. 'And be careful. Other people are looking for him. Other people are looking for this.' She handed the key to Jaq.

It was a silver key with a bottle-opener-shaped bow, the shaft cut from a half-cylinder with three-dimensional grooves and a number – 12016834 – engraved on the smooth side, a threaded hole bored sideways through the shoulder, a triangular opening with the word ABLOY engraved diagonally and an embossed red spot.

Jaq held it up to the light. 'Elena, where is Sergei's locker?'

'I don't know exactly.' Elena dropped her eyes and turned away. 'Somewhere in the zone of alienation.'

Chernobyl.

Tuesday 31 May, Kiev, Ukraine

Frank paced up and down the arrivals hall of Kiev International airport and glowered. Where was The Spider, damn him? What was the point of accountants if not to be boringly predictable and completely reliable?

The flight had been delayed. The excuses grew increasingly pathetic: fog, wind, storms, late inbound, no plane, no crew – no fucking idea how to run an airline, more like.

He flung his bag to the floor in front of the meeting point and winced at the pain in his hip. A bumpy flight due to turbulence and incompetence. Business class meant nothing more than curtaining off two rows of economy seats at the front of the plane – cramped and uncomfortable. The service was a disgrace, no refreshment was offered until an hour after take-off. The gurning air hostess had the body of a professional weightlifter; the selection of drinks was dismal and the food inedible. Why the hell had he come here, anyway? Frank scanned the people milling around in arrivals – short, squat, poorly dressed. What a dump. Where was Pauk? It was not as if The Spider was easy to miss. Frank dialled his number. No reply. It went to voicemail and Frank cut the call without leaving a message.

A clutch of uniformed drivers with placards stood near the door, the names unintelligible in Cyrillic script. At least the symbols for the international hotel chains were recognisable. Frank nodded at the man holding a sign with an emblem he recognised.

The driver frowned. 'Natasha Lominsky?' he said.

'Frank Good.' *Idiot.* 'I have a reservation.'

The driver pointed to the taxi rank outside.

Frank grabbed the cardboard sign from the driver and ripped it in two. 'I don't travel in public taxis in cities I don't know,' he said. 'Now take me to my hotel.'

The driver muttered something incomprehensible and pulled out his mobile phone. There was a hurried conversation. A young woman with a name badge on the lapel of her blazer marched towards them and the driver backed away, retrieving his torn sign.

'May I help you?'

This one spoke English at least. 'Yes, I require an immediate transfer to my hotel,' Frank said.

'Did you book a transfer?'

'Yes.' Not with the hotel, with Pauk, but since The Spider had failed to materialise, the hotel transfer would have to do. 'And this delay is unacceptable.'

'I am so sorry.' She gave him a long, cool look and then wrinkled her brow. 'Please come with me.'

By the time the hireling had sorted out a car and it was drawing up outside the hotel, night was falling. Frank checked in and tried Pauk's mobile again. This time he left a message.

'Listen, you bastard, when I get hold of you I am going to string you up and flay you alive.'

What the hell was he going to do in Kiev without the accountant?

Wednesday 1 June, Kiev, Ukraine

Jaq parted the curtains and gazed out over the city. A trickle of traffic filtered through the wide street from the medieval St Sophia to the baroque St Michael's Church, linked by the statue of a Cossack on a rearing horse. The early-morning sunlight glinted on golden domes; Kiev was more beautiful than she'd expected.

An old woman, dressed entirely in black, trundled a wooden cart to the corner of a public park and began to set out a stall. She unpacked a flimsy trestle table and stacked it with huge green spheres until it sagged in the middle. The woman drew a cleaver from her cart and split the nearest fruit, cutting it into slices of red flesh dotted with black seeds. Watermelon. A pang of hunger stabbed at Jaq's belly.

Jaq investigated the breakfast room. Grand, high windows and billowing white curtains, but the food on offer looked tired and unappetising: instant coffee and long-life milk; limp, fatty bacon; sausages too pink to contain much meat; scrambled eggs polymerised to rubber; bread like cotton wool; croissants that stuck to the roof of your mouth – you didn't need to try them to know. Instead she wandered out into the street, straight over to the watermelon stall.

'*Dober den.*' Jaq handed over a note in exchange for a slice of watermelon. As she bit into the flesh, a warm burst of sweet juice dribbled down her chin. The old woman handed her a paper napkin before counting out the change in hryvnia: *Odine, dva, tree, cheteerye, piyat, shist, sim . . .*

The meeting point for the tour was a hotel in the centre of the city, an American-designed steel and glass building towering over the zigzag of tiered Soviet blocks. Inside, it looked like any other business hotel anywhere in the world. A high-ceilinged atrium with bright lights and carpeted floor, men in suits coming and going. A group of people in more casual dress gravitated to a man with a clipboard: order from chaos. She observed him for a moment. Medium height, slight build, light brown hair, mid-thirties? He looked up, scanning the atrium, fixing on her approach. Warm, intelligent eyes.

'Is this the Chernobyl tour group?' she asked.

'*Chor*nobyl.' He smiled. 'Chernobyl is the Russian name. In Ukrainian it's *Chor*nobyl.' He ruffled through the papers on his clipboard and found the form with her photograph. 'Ms da Silva?'

'Just Jaq, please.'

They shook hands. His grip was warm and firm.

'My name is Petr. I am your guide. Your passport, please.'

As Jaq swivelled to reach into her backpack, she caught a sudden movement from the corner of her eye. A man hurried towards the lift behind her, his face turned away. Something about his purposeful stride was familiar. As the lift doors opened, he pushed his way in before the occupants could leave. She waited for him to turn and face the lobby; but he remained with his back to her until the lift doors closed. Who did she know in Kiev?

No one.

Petr interrupted her train of thought, introducing her to the rest of the party and inviting them all to board the coach, thirteen ill-assorted tourists: a cheerful American couple and their sullen adult daughter, two silent Norwegians laden with filming equipment, a group of four French students, a Chinese couple and a pale, red-haired English hippy.

The modern coach sped through nondescript suburbs, row after row of concrete blocks of flats, following the line of the River Dnieper to the north and into the zone of compulsory resettlement.

The forbidden zone.

Jaq rested her head against the window. Ironic. Approaching

the contaminated land with a bus full of strangers, she'd never felt safer. Over the last few weeks, escaping from Slovenia, surviving Belarus and preparing for the Ukraine, she expected the police to catch up with her. The zone of alienation around the Chornobyl power plant was the last place anyone would be looking. For the first time in days, she could relax. Jaq slumped back and closed her eyes.

Wednesday 1 June, Chornobyl Exclusion Zone, Ukraine

The punishment cell in the complex stank of blood and piss and shit. Boris didn't know what was worse, the smell of his own blood and piss and shit, or the knowledge that it mingled with effluvia from others tortured here before him. All feeling had gone from his hands, wrists, arms and shoulders, stretched by his body weight dangling from the manacles.

For the first time in days, his mind was clear. Yet now he almost wished for the hallucinations to return. Anything was preferable to reality. He groaned as a creak of hinges and crack of light announced Mario's return. The short, square man standing in the doorway held a filleting knife against a whetstone. The blade gleamed in the dim light, sparks flying from the edge, illuminating his swarthy face long before he flicked the light switch. Not a man, but a monster.

Mario lowered the chain hoist until Boris's feet touched the filthy prison floor. He ran the flat side of the knife across Boris's cheek, turning it at the last minute to nick his nose. 'You know the punishment for failure?'

'I can explain.' It took enormous effort, but Boris kept his voice calm and even. 'Don't kill me.'

Mario smiled. 'By the time I have finished, you'll be begging me to kill you.'

'No.' Boris took a deep breath. 'You need me.'

Mario laughed. 'No one is indispensable. Least of all an assassin who fails to deliver.'

That stung almost as much as the cut to his nose. Boris was a chemist. The other stuff was more of a hobby. He swallowed hard. 'To find the tracker, I need Silver alive. That's why I didn't kill her.'

Would Mario swallow the excuse? The idea had come to him afterwards, escaping from the mess in Kranjskabel. Why do all the hard work himself? Why not let Silver do it for him?

Mario paused. His eyes glinted. 'What do you mean?'

'She'll find the tracker for us.'

'And when she does?'

'I'll take it from her.'

Mario laughed, a long, slow sneer. 'Your track record with Silver isn't great. The Spider will send someone else.'

Boris sighed and closed his eyes. 'This job requires finesse.' If she could outwit him, surely she could outwit anyone? 'You need me.'

'Indeed. We have a little testing programme planned for you.'

More nerve agents. Boris the human guinea pig.

'Wait.' He swung round, thinking fast. 'You need me to unlock the tracker. To disable it. I'm the only one who knows how. Otherwise you'll send out a beacon to all the enforcement agencies in the West, and they'll wipe you out.'

The lie stopped Mario in his tracks.

'Did you tell The Spider?'

'I didn't get the chance.' Not before being served the Moscow Mule with added kick.

Mario hesitated. 'What if she doesn't find it?'

'She'll find it.' Boris sent up a silent prayer that it would be so. Don't let me down, Silver.

Mario sheathed the knife.

'Wait here,' he said.

Like he had a choice.

Wednesday 1 June, Kiev, Ukraine

'Everybody out!'

Jaq woke with a jump, disorientated, confused. The purr of the diesel engine died away with a final wheeze and splutter. Outside, a white two-storey building stood sentinel over a red barrier. She could just make out the sign: Дитятки. She mouthed the letters, like breaking a code, *Dityatki*. The first checkpoint. The edge of the thirty-kilometre exclusion area.

She had the key. Now to unlock the secret. A secret worth killing for. A lurch in her stomach spread into a shiver across her shoulders.

Descending from the bus, the group waited in two lines to have their bags and papers checked. Jaq chose the shorter queue, but it didn't seem to be moving. The strident tones of the Americans echoed through the hall. The French students switched to the other line and Jaq moved to the front where she could see the American girl involved in a tug of war with a Ukrainian security guard, each with a strap of her satchel.

Petr rushed forward. 'No food or drink allowed.'

'That's not fair!' The American girl stamped her foot.

'We're coming back this way,' Petr said. 'You can leave your bag in one of the lockers.'

Lockers. Could it be this easy? Jaq moved to shadow them.

'Wait, please.'

The guard pointed at Jaq's bag. She opened it and the guard waved her through. She followed Petr to a small anteroom filled with banks of modern lockers. All with combination locks. Not a single keyhole.

Damn.

'Totally ridiculous,' snarled the girl as she unpacked three cans of Diet Coke, two bags of crisps, an open packet of chocolate chip cookies and a family-size box of Reese's Peanut Butter Cups. 'I'm not eating any of their stupid food. Why can't I bring my own?'

'Because this is a contaminated zone. Radioactive particles could settle on your snacks,' Jaq said. 'And if you ingest radionuclides, it will make you very sick indeed. First you vomit, then your hair falls out.'

Petr flashed her a grateful smile.

'Who asked you?' The girl slammed the locker door and flounced back to the queue.

The group passed through the security fence and a bus drew up. Unlike the modern Pullman coach from Kiev, the tour bus had seen better days. The windows were covered in fine red dust, and splashes of mud fanned out from the wheel arches. As she stepped inside, the smell made Jaq recoil: a mix of body odour, cigarette smoke and spilled diesel. She sat near the front and closed her eyes. Deep breaths.

'Your attention, please!' Petr stood at the front of the bus and clapped his hands. 'We are going to show you a video.'

The TV screen flickered into life and music blared into the bus.

'*On the night of April 26, 1986, an explosion destroyed Reactor Number Four of the Chornobyl nuclear power plant. Thanks to the bravery of the engineers, firemen and liquidators, a complete reactor meltdown was avoided. More than 100,000 people were evacuated. Work on partially constructed Reactors Five and Six was halted. The other reactors – One, Two and Three – continued to operate until the year 2000, when they were also shut down. This area will continue to be contaminated for 200,000 years.*'

'Is it safe?' the American girl asked, glancing across at Jaq.

'Don't worry, it's safe,' her father replied. 'Otherwise they wouldn't let us in.'

Safe? What is safe? The bitter irony of the Chernobyl accident – it happened during a safety test.

Jaq taught undergraduate courses on safety at Teesside

University as a visiting industrial lecturer. Chernobyl was a case study. In the rush to meet energy targets, the safety tests due to be completed before start-up were bypassed. The safety engineers insisted the tests were carried out anyway. On a live plant.

Nuclear fission – the splitting of atoms – produces heat. The heat is used to make steam. The steam is used to drive turbines to create electricity. The electricity is also used to pump the cooling water required to control the reaction. During a reactor shutdown, water is still required, but the electricity to run the pumps is lost. So backup diesel generators start automatically. Tests on Reactor Four showed that the backup system took over sixty seconds to reach the required power. Far too long for the reactor core to remain without cooling.

Someone had the bright idea that the combination of steam turbines coasting down and diesel turbines winding up might provide enough power to run the emergency cooling pumps, elegantly bridging the power gap.

Previous tests had proved unsuccessful, but a fourth test was scheduled for 25 April 1986, in advance of a planned shutdown on Reactor Four.

To allow the test to be repeated if it failed the first time, most of the reactor emergency shutdown systems were disabled.

Brilliant.

Like airplane pilots ordered to experiment with the engines in flight.

Jaq closed her eyes and sighed.

The video continued.

'*The core is still active, and the original sarcophagus is crumbling. We are building a new shelter for Reactor Four: the New Safe Confinement. Height 100 metres, 165 metres long and 260 metres wide, it weighs 30,000 tonnes. It will have a lifetime of more than 100 years.*'

At the Leliv checkpoint, the edge of the ten-kilometre exclusion area, they halted for another security check and safety briefing. Jaq stood and stared out at the cranes around the New Safe

Confinement towering over the trees, ready to build an arch big enough to cover St Paul's Cathedral and the Tees Transporter Bridge with room to spare.

Back on the bus, Petr reinforced the safety instructions, his voice calm and steady, woven with gravitas. Don't eat or drink anything unless given to you at the final stop. Don't touch anything. Don't put anything down on the ground, don't pick anything up. Do not, repeat do not, take anything from the zone.

Was he staring directly at her? Jaq waited until he'd turned away, then allowed her hand to stray to her neck. The key hung on a ribbon, nestling unobtrusively between her breasts, warm from her skin. Her fingers stroked the shaft – one curved edge and one flat – and followed the grooves cut into the metal. The texture changed at the top where a triangular hole opened out and the ribbon was threaded through. Smooth, polished metal, with the maker's name engraved beside an embossed circle and the number. A number she had memorised: 12016834.

The zone of alienation was 2,600 square kilometres. How could she search an area so vast? How to find the right locker for this key?

Wednesday 1 June, Chornobyl Exclusion Zone, Ukraine

Petr handed out yellow boxes, about the size of a mobile phone but thicker and heavier. 'We are now entering the ten-kilometre exclusion area. Wear this personal dosimeter at all times. Switch it on like this.' He held one up and demonstrated. 'Test it like this.' Throughout the minibus the sound of bleeping rang out.

The Chinese couple whispered behind her.

Petr continued. 'Your personal dosimeters are set to alarm when they detect higher than normal background radiation.' He smiled. 'When the alarm goes off, do not worry. If you follow my rules, your total radiation exposure today should be less than you would be exposed to flying from here to America.'

The red-haired woman leant across the aisle towards Jaq. 'That's why I never fly,' she said. 'I'm Megan, by the way.' She extended a hand. The smell of patchouli made Jaq queasy; she turned her head away as she shook hands. Her stomach rumbled and lurched for the rest of the drive. When the bus stopped and the door opened, she was the first one out.

The ghost town of Pripyat stood in the sunshine, block after block of flats, all deserted and eerily silent. Unchanged for almost thirty years, the town remained frozen in time. Petr led the group along Lenin Street to the main square, pointing out the Palace of Culture, the Polissya hotel and a vast supermarket, still stocked with tins, packets and boxes from 1986. A Ferris wheel, listing at a slight angle, towered over what had once been an amusement park.

Jaq ignored the cramps in her stomach and followed the group into the old sports centre, marvelling at the Olympic-sized swimming pool, now a huge void surrounded by broken tiles and windows.

Swimming pool. Changing rooms. Locker room. Where? Jaq looked around and spotted two openings at the far end of the pool. Each had a tiled square footbath. One for men and one for women. Jaq fell behind as the tour group exited to the gymnasium. She crunched over broken glass and tiles and slipped through the nearest of the two openings, ducking under a fallen beam. This area, away from the official tour route, was far more dilapidated and she moved carefully, looking up to see daylight where the ceiling had fallen in. The separating wall had crumbled. From the women's area, identifiable by the lack of urinals, she could see into the men's. Both had toilets, showers, benches and . . . yes! . . . banks and banks of lockers.

The wrong type.

Rusting metal doors with three horizontal vents, and simple locks opened by a tiny flat key. Nothing that matched the monster key around her neck. Heavier and heavier, it dragged her down as she walked past each row, checking every locker, until finally after six hundred she had to admit defeat. Hopeless. Why had she come? The stomach pains were becoming more insistent. More frequent cramps. Damn it. She couldn't afford to be ill. She gritted her teeth and hurried back to rejoin the group milling around in the gymnasium. Petr scrutinised her face and raised an eyebrow.

'Everything okay?' he asked.

She nodded. If only he knew how far from okay things really were. Could she show him the key? Risk asking him where to find the locker it opened? And try to explain how she had come by it?

'Keep up, I'd rather not lose you.' He held her gaze for a little longer than was comfortable. Did he suspect something, or was he flirting? His voice so warm, his eyes so kind. Another time, another place, she might have explored that tenderness. Right now, she had enough on her plate. She looked away.

They made their way through deserted residential streets. An eerie silence, a post-apocalyptic hush. Unsettling.

'Creepsville,' the American girl said. 'I'm not going any further.'

Her father frowned. 'Don't you want to see where Svetlana grew up?'

'Boring.' The girl sniffed and snapped on her headphones.

Jaq turned to the American woman. 'You lived here?'

'I was born in Pripyat.' She had no accent, but she spoke slowly as if it caused her some pain. 'My parents worked at the power plant. We were evacuated the day after the accident. They said it was temporary, but . . .' Her voice trailed away for a moment.

'This is the first time she's been back,' the American man said. 'I'm Brad, by the way.' He held out a hand and Jaq shook it. 'This is my daughter, Pip,' he indicated the sulky girl, 'and my wife, Svetlana. I wanted to share this moment with her.' He kissed his wife on the lips, and his daughter made a disgusted face.

Jaq walked on, leaving them to their private moment, but Svetlana caught up with her.

'There.' Svetlana pointed. 'That's where we lived.'

She indicated a featureless concrete block of flats, shading her eyes as she counted up to the eleventh floor. 'My mother loved it here,' she said. 'We had space to live. A bedroom each. A modern kitchen: a big fridge, a washing machine, endless hot water. The shops were full of things: meat and vegetables and fruit. No queues. And outside,' she swept a hand across the tangled overgrowth, 'parks with roses. Roses everywhere. My mother loved roses.'

'Let's go see.' Brad moved towards the block of flats.

Petr rushed over to intercept them. 'Keep to the path. It's been decontaminated out here, but inside there are radioactive hotspots. Anyway, the buildings have not been maintained, some are no longer structurally sound.' He nodded at the warning signs.

The group traipsed towards the lake where their bus waited for them.

A dog howled. Svetlana started talking in a low, dreamlike

voice. 'We cycled out for a picnic – here, at the lake, to watch the firemen. It was such a warm weekend. I thought it was exciting, the beautiful red glow in the sky.'

'Your parents let you?'

'They didn't know how dangerous it was. No one knew.'

'Some knew, but said nothing,' Brad said.

'The next day, the buses came for us. We couldn't take much, but they told us it was only for three days. We had to leave Sasha, our dog, behind. They told us they'd take care of him, there was nothing to worry about.' She wiped a tear from her eyes. 'He ran after the bus. He didn't understand why we were leaving him. He ran down the road for miles and miles. Barking. All the way. I was crying.' She was weeping openly now. 'Crying for Sasha.'

Jaq followed her gaze. A long, straight road. Conifers on either side. Her stomach twisted.

'I promised Sasha. I promised to come back for him.'

Brad held her hand. 'You were a child, Svetty. You weren't allowed to go back. No one was allowed. Your parents did what they had to do.'

'I keep thinking I hear Sasha.'

'They shot all the animals,' Pip said. Svetlana winced and Brad glared at her. 'They had to, didn't they? They could have wandered out of the zone. Spread the radiation.'

'Sweetheart, you're not helping your mother—'

'She's not my mother!' the girl hissed.

Jaq backed away. Megan followed her. 'She's right, though,' the pale woman whispered. 'What did the animals do to deserve it? It wasn't their fault.'

Jaq said nothing. What could she say? When terrible things happen, the vulnerable suffer most: the elderly, the sick, pregnant women, babies, children, animals.

Megan wheeled in front of Jaq, forcing her to stop. Jaq narrowed her eyes. There was something unsettling about Megan up close. Red hair scraped back into a bun so tight it stretched her skin and made her watery green eyes protrude from their red rims. Her pale skin was dotted with freckles. Her choice of clothes – unbleached cotton shirt, hessian pantaloons and flat leather brogues – contrasted oddly with the bright lipstick, a purple slash across her thin lips. A large silver cross on a long chain bounced as she moved.

'Do you know what this is?' Megan pointed at a clump of weeds growing out of a tree stump. She didn't wait for a reply. 'Wormwood.'

An unremarkable plant, a small bush with red stalks, light green fronds and a cascade of yellow flowers. Megan bent forward, as if to pluck a flower.

Jaq extended a hand to stop her. 'Best not to touch—'

Megan grabbed Jaq's hand. 'Wormwood,' she repeated, and brought her face so close Jaq could see the blood vessels in her eyes. 'The Ukrainians call it *chornobyl*. All this,' she waved at the remains of the power plant in the distance, 'this catastrophe was predicted in the Bible.'

Oh, great, just great. Stuck in a post-apocalyptic radioactive hellhole, on a hopeless search for a locker, with a cheerleader for the God squad. And her stomach hurt like hell. Could her day get any better?

'Book of Revelation, chapter eight.' Megan's voice changed as she opened her mouth wide to declaim in a monotone. '"And the third angel sounded . . ."'

The halitosis from the woman's mouth made Jaq gag and she tried to pull away, but Megan tightened the grip on her hand.

'". . . a great star fell from heaven burning . . . upon the fountains of waters . . ."'

The queasy feeling in Jaq's stomach had taken on a life of its own. This time she wrenched free and stepped back. The air around her turned white, a high-pitched whine thrummed in her ears and bile rushed into her mouth.

Megan raised both her hands into the air. '"And the name of the star is called Wormwood . . . and . . . many men died of the waters, because they were made bitter".'

Jaq collapsed onto the tree stump beside Megan and bent her head.

'Hey, you can't sit there . . .' Petr came running.

Jaq clutched her stomach and groaned. 'I'm going to be sick.'

'What is it?' Brad rushed towards Petr. 'What's wrong with her?'

The waves of nausea accelerated until she could no longer control the spasms, and Jaq turned her head away and vomited. Again and again.

Petr ushered the group towards the bus and waited beside her until the retching stopped.

'Are you okay?' Petr offered a handkerchief. 'Can you make it back in the bus?'

Pip glared at her father. 'She's not coming with us, is she?'

Jaq wiped her mouth and looked up at the group hovering by the steps of the bus. 'No.' She shook her head. 'I'm going to be sick again.' And worse.

'Radiation sickness!' Pip shrieked. 'She's just the first. We've been to all the same places. We've all caught it. We have to get out of here!'

Petr bent down and lowered his voice. 'Did you eat anything?'

Jaq shook her head. 'Nothing.' Nothing since breakfast. Not even breakfast. Just the slice of watermelon. At the thought of it, her stomach heaved again.

'"And the name of the star is called Wormwood"!' Megan screeched and moved towards them.

Petr intercepted Megan, standing between her and Jaq. 'Please calm down and get on the bus.'

Megan dodged round him and ran back to the patch of wormwood. '"And the waters were made bitter"!' She began twirling, round and round. Her hair came loose from the clasp and red curls fanned out around her face as her voice grew louder and louder. '"And many men died of the waters"!'

The American girl screamed.

'I'm calling for help,' Petr said.

Jaq vomited again. Other members of the tour group began to wail.

'"And the name of the star is called Wormwood"!'

The shouts became louder as Brad pushed his daughter back into the bus.

'"And the name of the star is called Chernobyl"!'

'Jaq.' Petr returned and bent down beside her. A warm hand on her shaking shoulder. 'I'm sorry, but I have to get the group out of here. Do you understand?'

Jaq nodded and waved him away.

'Stay here. Don't move. The medical team are sending an ambulance for you.' He hesitated. 'I'd rather stay, but . . .'

'Go!' Jaq managed a wan smile. 'Just take the lunatic with you!'

Jaq wasn't sure how Petr loaded everyone back on the bus, but when it trundled away and the wailing and shouting finally faded, the silence was glorious. Now that her stomach was empty, the pain in her abdomen abated. The sun warmed the back of her neck and she stopped shivering as she turned west to face it. A giant ball of radioactivity up there in the sky. A huge atom bomb. The energy source of all life on earth.

Was she suffering radiation sickness? Unlikely. Even the men who received fatal doses on the night of the reactor explosion didn't display the symptoms until hours or days later. Much more likely to be something she ate in Kiev. Her breath came more slowly and her clenched muscles relaxed. The silence was not so silent now. Nature had taken over, laying on a concerto especially for her.

First came the smaller insects, beating wings, a low background hum. Then the bees, the buzz of activity as they danced directions for the hive. The flap of a bird's wing followed by a splash as it plunged into the lake, the massive cooling pond created for the nuclear power plant. The rustle of long grasses. A little mammal – a shrew? No, bigger – a beaver. The chatter of a woodpecker and the song of a thrush. The neigh of a miniature horse lapping water with a long pink tongue at the edge of the lake, tossing its silver mane as a foal joined it.

The howl of an animal. A dog? The ghost of Sasha? Or a wolf.

She vomited again.

Where was the ambulance?

Wednesday 1 June, Chornobyl Exclusion Zone, Ukraine

The medical centre, assembled from a set of lead-lined shipping containers, was kitted out with every modern medical device. The walls inside were smooth and white, bright halogen lights recessed into the ceiling giving the interior the feel of a spaceship. Jaq lay on the stretcher and shaded her eyes.

The ambulance staff had arrived in Pripyat shortly after Petr and Megan and the rest of the tour party left. Just before the wolves. Despite her protests, the paramedics insisted on wrapping her in a foil blanket, strapping her onto a stretcher and inserting a drip into her arm to feed a cocktail of fluids directly into her veins. She was feeling quite disorientated by the time they handed her over at the decontamination suite.

Bleep . . . bleep-bleep . . . bleep. A nurse scanned her with a portable instrument. 'Nothing to worry about,' he assured her. 'Quite normal, low even. Whatever made you sick, it's not the radiation.'

He showed her the readings. Liquid crystal display on a silver screen. Numbers, just numbers. What was in a number? After the accident, the managers at Reactor Number Four reassured Moscow about the radiation. While high, it had not exceeded four. Their instruments only went up to four. When they brought in a wider-range instrument and it too went off the scale, they decided it was broken. They couldn't believe the readings. Man's ability to fool himself is legendary. Woman's, too. Why had she come here? Her hand strayed to the key around her neck.

The doctor arrived, a gruff Finnish woman with a blonde bob wearing a white paper suit. She asked Jaq questions – what had she done with the tour group, when had she last eaten, what had she eaten.

'Aha, watermelon!' she exclaimed when Jaq told her about the street seller in the park. 'Never touch watermelon here. It's a giant sponge, sucks up dirty water and barely filters it. Any bacteria remain in the flesh.' She shook her head. 'Radionuclides, too. We'd better err on the safe side and check out all possibilities.' She explained the tests.

'When was your last period?'

Jaq counted back. 'Ten days ago.' Regular as the moon.

'Is there any possibility that you are pregnant?'

Jaq shook her head. 'Absolutely not.'

'Undress, please.'

Jaq shivered as she removed her clothes and slipped into a white paper gown.

'That key as well.' The doctor pointed to the yellow ribbon round her neck.

Jaq hesitated. After all this effort she didn't want to be parted from the key. She looked up at the doctor, realised she had no choice and reluctantly handed it over.

Once the doctor had finished the examination and samples had been dispatched to the lab, she explained how the emergency medical centre had been pre-assembled and brought in for the workers on the New Safe Confinement project. The medical staff monitored and controlled the exposure to radiation, but they also dealt quickly with the conventional injuries of a big construction site; slips, trips and falls had the potential to be much more serious in such a contaminated area.

When the first test results came back from the lab they indicated *Staphylococcus aureus*: food poisoning.

'The worst is probably over. We'll need to keep you hydrated for twelve hours, but not here.' The doctor handed Jaq the tray containing her belongings and then stopped to peer at the key. 'I thought you were with the tour group,' she said.

Jaq's scrutinised her face, trying to read the reason for the reaction. 'Day off,' she said. 'I'm an engineer.' As if it explained anything.

'They didn't tell me.' The doctor scratched her head. 'Have you been here long?'

'Just arrived.'

'But you left something in Sector Twelve?'

Sector Twelve? Did the number on the key contain some sort of location code? Jaq waited.

'I wonder why they put you in Sector Twelve, I thought it was empty.' The doctor stood and inspected a large map on the wall. She started at Reactor Number Four and followed a thin line, highlighted in yellow. The line of the steel rails which would be used to carry the giant sarcophagus – the new containment for the core – and slide it over the stricken reactor. Her finger continued back to the New Safe Confinement construction site and the medical centre and then moved north-west, almost to the border with Belarus, where a workers' dormitory was shown, labelled with the number twelve. 'Well, it's a good place for isolation and there's a medical room. Do you want to go back to Sector Twelve?'

The urge to vomit gripped Jaq. She barely had time to agree before her stomach emptied itself into the basin, acid burning her throat and nose.

'Okay, Sector Twelve it is.' The doctor called for the nurse. 'Let me know when you feel strong enough to move, and I'll arrange the transport.'

If Jaq hadn't felt so wretched, she would have cheered.

Wednesday 1 June, Chornobyl Exclusion Zone, Ukraine

Night was falling as the ambulance turned off a straight forest road and stopped in front of a wooden building. The dormitory in Sector Twelve appeared far from deserted: lights on and faces at every window. If anything, it looked positively overcrowded.

It seemed that others thought so, too. Jaq waited in the ambulance while an argument raged in front of the headlights. The superintendent of the dormitory had the stance of a formidable woman – short and square with a lofty blonde beehive, hands on hips, legs planted wide apart. She couldn't hear what was said, but the body language was eloquent. *Minha Nossa Senhora.* Any moment now she'd be found out. The superintendent would deny all knowledge of her, demand to know why she was impersonating a worker on the New Safe Confinement project, how she'd come by a locker key. The real workers would denounce her.

Dimitri, the nurse who accompanied Jaq in the ambulance, was not, however, a man who took no for an answer. He'd been told to set Jaq up in the medical room of Sector Twelve with a rehydration drip, and he intended to obey orders. And he prevailed.

As Jaq followed him into the little room, the hostile stares of women drilled into her back. 'Do you know any of these people?' Dimitri asked.

'I'm new,' Jaq said.

At this the superintendent turned and inspected her with something approaching a smile before she closed the door.

'Rather you than me,' Dimitri said. 'This place gives me the creeps.' He set up a new drip. 'Get some rest. This will last twelve hours. Only water for another twelve. I suggest you get a thorough medical back in Kiev.'

Jaq dozed. Chatter. Shouts. A bell ringing. Clattering of glass and china. Footsteps. Doors closing.

As the voices faded away, the noises of the night grew louder. Wind in the trees, pitter-patter of rain on a plastic roof. Creaks and juddering as the building cooled and contracted.

Her stomach muscles ached, but it was the insistent pounding of her heartbeat that kept her awake. She checked her watch; several hours had passed since she last vomited. Could she risk exploring the building? The Finnish doctor didn't even have to look at the number to know the key was for a locker in Sector Twelve. Why? Abloy was a Finnish lock company. The massive construction project – New Safe Confinement – was made up of several consortia: French, Italian, German, Swedish, Norwegian . . . why not Finnish, too? The door and windows were smooth blond wood with stainless steel fittings, a clean, modern design. A modular construction built in the safety of a factory and then shipped in pre-assembled modules to the contaminated zone.

Sergei had left Snow Science for a job in the Ukraine. The New Safe Confinement project? He would be a perfect candidate – an engineer, native tongue Ukrainian, familiar with the Chornobyl complex.

Jaq crept to the door of the medical room, the portable drip trailing behind. She pressed an ear against the door. All quiet. Her legs wobbled. Food poisoning? Trepidation? No – excitement. The key seared her skin. The locker must be nearby. She was sure of it.

Her heart jumped to her mouth as she opened the door and peered out. One corridor led back to the main door – the way she'd come in – and another branched off at right angles into darkness. Where was the locker room?

Jaq padded into the corridor. The liquid sloshed in the bag as she lifted the drip stand wheels over the door seal. Her bare feet made a slicking noise on the grey linoleum of the corridor. She stopped before each intersection to listen and took a left turn at the first crossroads.

Stop. Laughter drifting down the corridor. Someone coming. Hide. But where? Jaq crouched in the shadow of a stairwell, hoping the gleaming metal drip stand was not too obvious. Two women strolled past hand in hand carrying towels and shower bags. Jaq held her breath, exhaling only when they turned the next corner. At the sound of running water she followed the route they had taken. Water meant showers. Showers meant changing rooms. Changing rooms meant lockers. She found a new stairwell to hide in. Her legs began to shake. She longed to sit or to lie down, but she risked wrenching the drip from her arm if she tried. She couldn't even lean against the wall, the drip stand prevented her from moving fully under the stairs. Each cramped minute felt like an hour. When the two women returned, Jaq caught a faint whiff of apple as they passed. Once the corridor fell silent again, Jaq resumed her search for the amenities, making two wrong turns before she found it.

The changing room door had a large number five stencilled on the front. It opened on to a tiled room. A series of wooden benches gave way to a row of showers. She clutched her stomach as it heaved in response to the wall of smell – synthetic apple: pentyl pentanoate – and found a toilet. Individual stalls with doors. On the far side of the benches opposite the showers – a bank of lockers. Rows and rows of them. With the red spot and the magic word. Abloy.

But none with the number 12016834.

Damn.

Jaq crept out and crossed the corridor. Immediately opposite was another changing room with the number six. The same layout but with urinals. This one smelt different. Musty. Mouldy. Disused. Were there no men here in Sector Twelve? She moved past the toilets. She found the showers and benches and . . . bingo! Lockers.

Eight rows of forty lockers. She headed for the last line. The numbering on the key was logical. 12016834 – Sector Twelve, Ground Floor 01, Amenity 6, Row 8. Now all she needed to find was locker 34.

Bolas!

Row eight only had thirty lockers. All the other rows had forty. But at the last row the hot water pipes entered the room and made a 90-degree bend, robbing the space.

Jaq collapsed onto a bench and put her head in her hands. She shivered, suddenly cold and completely exhausted. Why had she come? What was the point? She didn't have the energy to rage. Or to cry.

The aniseed smell from the urinals finally forced her to look up. Turn towards the exit. And then she saw them. Ten extra lockers tucked in behind the door. Numbers 830 to 840.

Jaq's fingers trembled as she put the key in the lock of 834. Come on. This must be it. The shaft slid in and turned easily. She pressed the handle and opened the door. Empty. After all this, after all she'd been through to get here, how could it be empty? Sergei, you bastard, why did you send me on this wild goose chase? She thrust her hand to the back and it made contact with something other than steel. Rough, pliable, rectangular: a cardboard shoebox. Slowly, she withdrew the box then checked behind it. Nothing else. Heavy, more than shoes in here. Whatever Sergei had been hiding, the secret was in this box.

A noise. Footsteps outside. Jaq jammed the box under her gown and stood in front of the locker, pushing it closed with an elbow.

The door opened. The superintendent strode in and scowled. She spoke in Ukrainian with a heavy accent. Jaq clutched at her stomach, bending forward to hide the shoebox as she moved towards one of the stalls.

The superintendent's eyes widened. She probably didn't fancy cleaning up after Jaq. She strode out, banging the door behind her.

Jaq flushed the toilet and ran the tap at the handbasin in case the superintendent was listening. The bag underneath the drip pouch was big enough to take the shoebox. Jaq stuffed it in and shuffled back to the door.

The superintendent was waiting.

'*Inzhener?*' she asked. Engineer. The Russian word had the proper derivation. From ingenuity, not from engine. No wonder the English undervalued their engineers, regarding them as mechanics rather than inventors.

'*Da.*' Jaq nodded.

'*Distsiplina?*' What sort?

No point in lying. '*Khimicheskaya.*' Chemical.

The superintendent beamed '*Khorosho.*' Good.

Why was it good? Were they short of chemical engineers for the New Safe Confinement project? More the preserve of structural engineers and physicists, but you never knew what sort of challenges a major project like this could throw up. At least the superintendent seemed friendlier now.

'*Kompanii?*' Fancy some company? Ah, a bit too friendly.

Jaq liked women. As friends. How different might her life have been if she had felt the same sexual desire for women as for men. One day she might give it a try. But not here and not now. Not in a modular Finnish hostel in a radioactive fallout zone in a hospital gown with a drip attached. Not with an intimidating woman sporting a blonde beehive cast from stone. Right now, Jaq was more interested in Sergei's shoebox than a sapphic adventure.

'*Nyet.*' Jaq reached her room. '*Mne nuzhno pospat.*' I need to sleep.

Jaq closed the door. When the squeaking of slippers across the linoleum had stopped, Jaq opened the door to confirm she'd really gone.

Only then did she dare to open the shoebox.

Wednesday 1 June, Kiev, Ukraine

Frank paced up and down, his room too small to work off the seething anger. What the fuck was The Spider playing at? The message from Pauk said: *Unavoidably delayed. Speak later.* What sort of a message was that? Not a word of apology, no explanation. When was 'later'? Today? Tomorrow? Next week? Who was paying whom? Not any more. That was that. Frank had always been straight with his employees. Firm and fair. Although fair depended on your point of view, always firm. No one received a second chance to fail. Exactly what he told the project team in Smolensk. You fuck up, and you're out.

The internet was so slow it was useless, the mobile reception even worse. When he had complained, the dumb blonde at reception had suggested a tour of the city instead. As if he had time. Her sidekick strongly recommended a day trip to Chernobyl, except he couldn't even pronounce it right. As if he fancied being irradiated. Frank had told him where to shove those ideas.

Well, Pauk had missed his chance. Whatever business opportunity The Spider was so keen to promote, he had failed at the first hurdle. First rule of business – never keep Frank Good waiting.

He contacted the airline from the hotel phone. There was a flight out this evening via Paris – heaven forbid, Charles de Gaulle was the worst airport in the world. An earlier flight via Amsterdam might be possible, if he hurried.

Frank travelled light: a good-quality three-piece wool suit, Italian leather shoes, spare shirts – professionally laundered and folded for travelling, several sets of cufflinks, two ties – daytime and evening, black socks, clean crotch-hugging trunks, shaving kit, deodorant, aftershave and condoms. Everything he wasn't wearing packed snugly into his four-wheeled carry-on suitcase.

He shut down his computer and slid it into the padded section of his case. Where was his passport? He'd presented it at check-in. The receptionist had spouted some nonsense about a broken photocopier and promised to send it up. Now he thought about it, had they ever returned it?

Frank telephoned reception but got the idiot bloke who had no idea that working in a hotel might involve the ability to speak something other than sodding Slav gibberish. He stormed down and made a beeline for the pretty receptionist. The airhead with plump cheeks and dimples. On her face, as well.

She forced a smile as he approached. God, he could teach her to do better. A little slap and tickle and she'd be grinning like a Cheshire cat, opening those pouting red lips for him. He stiffened at the thought.

'Can I help you, sir?'

Bad luck, honey. Next time. Not that there would be a next time. No, she'd missed her chance. 'I'm checking out.' He gave his name and room number. 'You have my passport.'

She went to the computer and frowned. 'Sir, there must be some mistake. Your room is booked for three nights.'

'No mistake,' he said. 'I'm on the next flight out.'

'Was everything all right, sir?'

'No.' Frank listed the various deficiencies in the hotel. She stopped writing them down halfway through his diatribe and a faraway look came into her eyes. This one needed more slap than tickle. 'Give me my passport,' he concluded.

He examined her bottom as she searched the pigeonholes, coming back empty-handed. 'I'm sorry, sir. It's not here. We must have returned it to you. Are you sure it is not in your room?'

He scanned her name badge – Irina – and licked his lips. 'Do you like your job here, Irina?'

She frowned. 'Yes, Mr Good.'

'Then it is a pity you are about to be fired, Irina.'

She bit her lip and looked around for support. 'Sir?'

'Unless you find my passport in the next two minutes, I will kick up such a stink, lodge such a detailed complaint, specifically about you, that you will be on the street without references and blacklisted from every hotel in Kiev.'

Tears welled up in her blue eyes. 'But, sir. I didn't take your passport. It's not my fault.'

'Oh, but it is,' he snarled. 'Right now, you represent the hotel group. I am holding you, and you alone, fully responsible.'

She blushed, her alabaster skin turning wine red. 'But, sir . . .'

'Your American masters always take immediate action. Sack first, ask questions later.' He flicked a finger at her. 'My passport. There's a good girl.'

She scuttled away and returned with her supervisor, the useless bloke who barely spoke English.

'Your passhport,' he said. 'Whiss Mr Pauk.'

Frank snapped to attention. 'Pauk is here?'

The man said something unintelligible and the bimbo translated. 'Yes, sir. He checked in before you arrived.'

Frank slammed his fist on the counter. 'Then where the fuck is he?'

'Message, sir.' The man handed over an envelope addressed to Mr Good.

Frank tore it open and his jaw dropped.

Cocktails at 7 p.m. Rooftop bar. See you then.
Pauk

PS You weren't thinking of leaving, were you?

Wednesday 1 June, Chornobyl Exclusion Zone, Ukraine

A silence settled over the workers' dormitory in Sector Twelve. Alone in the medical room, Jaq stared at the open shoebox. Inside lay a rectangular gadget. Matte black with a lilac sheen. A Geiger counter? Not one of the clunky old instruments from 1986. This device appeared new – brand new. A screen covered two thirds of the top surface and a keyboard occupied the other third. A toggle switch, a power socket and an empty USB port punctured the side with the manufacturer's symbol – a wheel of fortune – embossed on the front. Underneath that, the name TYCHE in large bold letters. *Tracker* in italics underneath the name. Scratched onto the front, four Cyrillic letters – ключ. Of course. *Klyutch*. Russian for key. The key had opened the door to another key. The new key was some sort of Geiger counter.

With a battery that was completely dead.

Jaq barely slept, drifting in and out of dreams, waking to the noise of birdsong. The intravenous fluid bottle was empty. She tore the drip out of her arm, wincing at the sharp pain, and ran to the bathroom. Insides still liquid, but no more vomiting. Progress. She dressed and called the travel agency from a payphone in the hall. Petr sounded guilty for having abandoned her, and expressed relief that she had been given the all-clear. If he was surprised to hear she was in a workers' dormitory in Sector Twelve, he didn't show it. He promised to send special transport so she could join a Japanese tour group on their way back after a three-day trip. She dressed and waited outside. Soft rain fell, and she sheltered under an awning until the noise of an approaching car beckoned.

The taciturn driver raced over the potholed roads through

acres of forest. She was glad there was nothing left in her stomach to bring up. They travelled south, leaving Sector Twelve, close to the border with Belarus, passing the deserted city of Pripyat and skirting the Chornobyl nuclear power plant. She gazed out at Reactor Four, deceptively peaceful in the sunshine, no visible sign of the sleeping dragon trapped inside. The taxi skirted the construction site for the New Safe Confinement and continued by the lake, down the Pripyat River, through the buried village of Kopachi to the checkpoint at Leliv.

'Da Silva?' A woman with twinkling brown eyes and aubergine-coloured hair was waiting for her. 'From Petr's group? How are you feeling?'

Lousy. 'Better, thanks.'

The new guide introduced herself as Katya and invited Jaq to join her group.

Jaq gripped her bag as she stood in line for the exit check. She adjusted her position, bringing the bag closer to her body until the sharp corners of the box pressed against her ribs. The weight of the Geiger counter made her shoulder sag. Would anyone notice? What were the rules? No souvenirs from Chornobyl. But this was no souvenir; this was a key. A key on which her freedom, her professional future, if not her life, depended.

A group of middle-aged Japanese men went first, all identically dressed in brand-new microfibre hiking gear, just one rebel with a ponytail and facial hair wearing a pink and green paisley shirt over jeans.

'*Nastupnyy bud'-laska*,' the female guard shouted. *Next, please.*

Jaq stood aside to let an elderly Japanese couple go ahead of her. The man was given the all-clear. The woman set off all the alarms. Katya handed a piece of paper to the guard and he nodded her through.

Jaq looked at Katya and raised an enquiring eyebrow.

'Doctor's note,' Katya whispered. 'Cancer treatment . . . remains in the body for a long time. Now you.'

Jaq bit her lip as she approached the square metal frame. She was expecting it, but she still jumped as the lights flashed and this time a siren sounded.

'Shoes and bag,' Katya said.

She removed her shoes. The guard opened her bag and scanned the items one by one. A female guard opened the box and started shouting.

Katya approached the guard and raised an eyebrow. 'Your personal Geiger counter?'

Jaq nodded.

The female guard, rosy-faced with fury, pointed at the box, shaking a fist at Katya.

'This is a proscribed instrument,' Katya said. 'You need permission.'

'I have permission.' In the face of bureaucracy, attack was the best form of defence. Confidence and authority, determination to match intimidation. 'I was with another group. We left the paperwork here. Don't tell me they lost it?'

The guard stood with her hands folded under her bosom and glared as Katya translated. The male guard picked up the box and carried it to the back office, closing the door. Perhaps Jaq had underestimated the Ukrainian response to determination.

'Hey!' Jaq shouted. 'That's mine.'

'I'll sort it out,' Katya whispered. 'Get back to the bus.'

Jaq paced up and down beside the coach while the rest of the party waited inside it. The engine rumbled into life as Katya emerged from the security building. Empty-handed.

'What about my Geiger counter?' Jaq said, barring her way to the bus.

'Confiscated.'

Jaq stood her ground. 'I'm not leaving without it.'

Katya brought her face close and hissed. 'I don't know where you got that thing.' She took Jaq's arm. 'But right now, for your own safety, we need to get out of here.'

An armed guard emerged from the building, striding towards them with a phone to his ear. He didn't look friendly. Jaq let Katya haul her onto the bus. The doors hissed closed and the bus rattled away from the checkpoint.

Leaving the Tyche tracker behind.

Thursday 2 June, Chornobyl Exclusion Zone, Ukraine

The bus turned onto a long, straight road. The security building disappeared from sight and with it the best hope of unlocking the mystery. So close. Jaq thumped the seat beside her with a clenched fist. 'We have to go back.'

'Don't push it,' Katya said. 'You were lucky not to get us both arrested.'

Jaq gripped her wrist. 'But I need it.'

'I can't help you,' Katya prised Jaq's fingers away. 'Petr will sort it . . .'

Of course. Petr would be able to resolve things.

'. . . if you had permission.'

Jaq put her head in her hands. Hopeless. It was all hopeless.

They turned towards the eponymous town of Chernobyl, now Chornobyl, a few miles to the south-east of Leliv.

Katya spoke into a microphone. 'Can I have your attention, please?'

'*Minasama* . . .' The young Japanese interpreter bowed and took another microphone.

'Our final stop before lunch . . .'

The Tyche tracker languished in an office in Leliv. How long would it stay there? Would they plug it in and charge it up? Check the data? Or just wipe it? One way or another, she had to get the tracker back before they tampered with it. Whatever it took.

The bus pulled to a halt outside a monument. Could she slip away? Easy. Hail a cab? Unlikely. The road was empty. Walk? They had only travelled a few miles. Possible. And then what? How was she to prevail over armed security guards? She glanced up at Katya. The tour guide was her only hope.

She followed the rest of the group from the bus as they gathered around the monument to the emergency responders.

'*Nan nin ga shibo . . .?*

'How many people died?'

'Thirty-one people died as a direct result of the accident,' Katya said. 'One man died immediately, killed by the explosion and forever buried in the rubble, one of a heart attack. Most of the rest were flown to Moscow to be treated in a specialist unit.' She paused. 'They could not be saved.'

Jaq looked up at the sky. A slow, cruel death. Unimaginable pain as the body succumbs to acute radiation exposure. The skin turns black and begins to fall off. The body swells up; internal membranes ulcerate. The vital organs shut down one by one; the body shrinks and desiccates. And only then, after days or weeks of intolerable pain, does death arrive.

The Chernobyl first responders suffered from raging thirst, but they could not drink, sudden hunger but they could not eat. Touch was forbidden; they had absorbed enough radiation to become a source of danger to everyone around them. Not that they could bear to be touched through their burning, raw skin. Somehow their relatives cared for them while the doctors attempted to heal them. All to no avail.

Katya continued. 'In the months after the accident, 600,000 volunteers – scientists, miners, construction workers, soldiers – were drafted in to limit the spread of radioactivity. Killing animals, harvesting crops and burying them along with the topsoil, covering up everything contaminated.'

The liquidators. Jaq closed her eyes. Not all were volunteers. Conscripts as well.

Katya finished, pointing up at the statue.

'This memorial is to all the heroes. The inscription says: To Those Who Saved the World.'

An elderly Japanese woman produced a small wreath from her handbag and laid it at the base of the monument. The group bowed their heads and stood in silence for a moment.

Jaq mourned for the men and women of Chernobyl Reactor Four.

The engineers were given an unstable beast in a complex cage and then, on the night of 26 April 1986, were ordered to unlock all the doors to see if the monster could escape.

It did.

Jaq sat apart from the rest of the group at lunch. She needed space to think. How had she ended up here? The single tourist hotel in a ruined town? She flinched at the clink of steel cutlery on china plates as the tour group tucked into borscht, and gagged at the smell of roasting meat from the gleaming modern kitchen. The food would do her no harm, all brought in from outside the zone of alienation, all checked for isotopes of caesium-137. She wasn't hungry.

She'd read about Chernobyl, written about it, lectured on it; now she had seen the harsh reality for herself.

The rush for nuclear power in the Soviet Union accelerated with the cold war. To compete with the West, the Soviets needed affordable power. The coal, oil and gas lay deep underground in the most remote regions, difficult to extract and transport to populated areas. So, when uranium was found in Kazakhstan, the wartime nuclear programme was reconfigured to include domestic power generation.

The technology recommended by the expert Soviet team was rejected. An inferior design was chosen, one which would be faster to implement.

But there were two critical design flaws.

In the military design chosen for the Chernobyl complex, the coolant was water. As the water boiled, the reactor power increased, releasing more heat and causing more water to vaporise, reducing the water level further in a vicious circle.

The experts understood this design flaw. The automatic control system was enhanced with secondary safety systems including an emergency shutdown button which inserted additional control rods.

But there was a second design flaw. As the control rods descended, they displaced water, so instead of reducing the power of the reactor, the power momentarily increased.

When the night shift took over just after midnight on Saturday, 26 April 1986, an insanely unsafe safety test was in progress on Chernobyl Reactor Number Four.

A supervisor intervened; he hit the emergency shutdown button. The control rods started to fall. The entry of the graphite tip of the boron control rod into an already unstable reactor was the final straw.

The power surged.

The first explosion happened seconds later, lifting the 1,000-tonne upper shield. A second, more powerful explosion followed. Lumps of fuel and graphite were ejected from the core, catching fire as they hit the air.

Compared with the atomic bombing of Hiroshima, four hundred times the amount of radioactive material was released from the Chernobyl reactor explosion. Unprotected workers received fatal doses in less than a minute, though it took them weeks to die. Radiation. Jaq shivered. Something chilling about not being able to see or touch or smell or hear the danger. Did it make them braver? Did the first responders know what they were walking into? Some did, and yet still they came.

The engineers on shift that night knew the risks. They understood that to stay was to die; knew they were being exposed to fatal doses of radiation. And yet they never faltered. Had they not acted, the accident would have been a hundred, a thousand times worse, and Europe would be a different place today.

The scientists who came later certainly knew; those who mapped the zone of alienation, walking across contaminated land with handheld Geiger counters, fully aware that the dose they received doing their vital job made a slow and horrible death inevitable.

But what about the liquidators, the ones sent to clean up afterwards, the ones who were given vodka and cigarettes instead of protective clothing and dosimeters, medals instead of medical screening? The ones who died slowly? Were still dying? The former Soviet Union was good at that. Throwing men and women into the path of monsters. Sacrificing individuals to save the greater humanity.

What had she achieved by coming here? Nothing. She had one task. Only one task. To find Sergei's locker and retrieve the evidence she needed. She'd had it in her hands, and then let it slip away.

Metal chair legs screeched against polished tiles. The tour group were on the move, but the guide was nowhere to be seen.

She had to give it one last try.

The tour guide sat in the bus, talking on the phone. And there beside her, on the front seat was . . . no, it couldn't be . . . a shoebox.

Katya scowled as she handed it over.

Jaq hooted with relief. 'How did you—'

'Petr says you owe him lunch.'

The group passed through the thirty-kilometre checkpoint without incident and changed buses. When they reached Kiev, a surge of optimism overcame Jaq as she waved goodbye to the group from Fukushima.

Thursday 2 June, Kiev, Ukraine

'Where the hell were you?' Frank's anger was tinged with relief to see The Spider standing – well, stooping – over a high table in the rooftop cocktail bar. *And where the fuck is my passport? Careful. Find out what he wants, or you'll never get out of this shithole.*

Frank grabbed a high stool and joined The Spider. Their eyes were at the same level. Pauk must be six foot seven or eight, although the awkward way he held himself brought him back down to a normal height. Fucking weirdo.

'Good evening, Frank.' Pauk held out a hand. He grimaced as Frank yanked it, forcing him to straighten up. 'I do apologise for not meeting you earlier. I had some loose ends to tie up.' He groaned and resumed his contortions as Frank released his hand.

Christ, what had he been thinking of to do business with this creep? The Spider had better have something good. 'I am a busy man. What is this business opportunity?'

'Let us start at the beginning, shall we?' Pauk said. 'What are you drinking?'

Frank snapped his fingers at the waiter and ordered a Bloody Mary. Pauk asked for the same – without the tomato juice.

Pauk plopped a sliver of lemon and a single ice cube into his vodka. 'The company that Zagrovyl purchased on my recommendation, Tyche, do you know what they do?' He passed the tongs to Frank.

'I know what they did. Chemical weapons disarmament.' Frank added three ice cubes and returned the tongs to the ice bucket. 'But all the chemical weapons are gone now.' The company formerly called Tyche had been merged into Zagrovyl. Quite profitably, as it turned out. With all the bleeding-heart, whale-cuddling greens, even conventional chemical waste disposal was an increasingly lucrative business.

'Indeed, but what if I were to tell you that there are still some customers for their original market?'

'If the price is right, why not?' Frank shook first Tabasco and then celery salt into his drink. 'Is there a new cache of chemical weapons in need of decommissioning?' Interesting. Tyche had special know-how. They could name their price. Especially if idiots like the United Nations were involved. Money for jam. He added Worcester sauce and tested his drink.

'Not exactly,' Pauk said. 'More a shortage.'

'A shortage of chemical weapons?' Frank added more Tabasco. 'I would bloody well hope so.'

Pauk nodded. 'As a result, there are people who will pay a high price for such materials, an astonishingly high price.'

'Yes, prison. That's far too high a price for me.' Frank sipped his drink and appraised The Spider. 'Come off it, Pauk, you know there are sanctions. We're talking about controlled chemicals – the authorities are all over us like a rash. Zagrovyl can't get around international rules.'

'No, but my associates can.'

'I'm going to pretend I didn't hear you say that.'

'Tell me, Frank.' Pauk opened a packet of cigarettes. 'Why do you view chemical weapons with such horror?'

Frank's jaw dropped. Was this guy for real?

'Death is ugly – is the medium by which it is delivered so important?' Pauk extracted a black and gold Sobranie. 'Is it any better to be stabbed or bludgeoned, shot or have bits blown off? How many gunshot wounds or shrapnel lacerations are immediately fatal? Victims bleed to death, or linger for days as agonising sepsis sets in.'

Pauk lit the cigarette with a gold lighter. 'What if a sophisticated chemical cocktail induces a glorious euphoria to ease the passing?

What if the end comes faster? Is swift and painless? Is this not a form of mercy?'

'Keep your voice down.' Frank scanned the room; the barman had moved to the far end to wipe tables, but still, you couldn't be too careful.

Pauk took a long puff and exhaled, making no effort to direct the smoke away from Frank. 'Perhaps you believe government misinformation about the rules of war. That armies send smart bombs, guided with pinpoint precision, and soldiers check the insignia of the enemy uniform before they fire. That the only honourable weapons are those with a line of intention – the sight of a barrel, angle of a mortar, point of a sword, blade of a machete, thick end of a club. Perhaps you still believe that sophisticated weapons can discriminate between enemy combatant and innocent bystander?'

Pauk shook his head.

'Is modern warfare really so cleanly delineated? Don't guerrilla fighters hide among civilians? How many innocents die in bombing raids? What of napalm, landmines, IEDs?'

Frank rolled his shoulders. Was he really having this conversation? Why had he ever thought Pauk smart? Until he got his passport back, he had no choice but to listen to this monster, to humour him.

'What if chemical weapons fall into the hands of terrorists?' he asked.

Pauk smiled. 'And what is a terrorist? Isn't one man's terrorist another man's freedom fighter? Anyone who plays by different rules of engagement? Anyone who targets the powerful, those who rely on their conventional armies to preserve the status quo? The patriarchy? Those who abuse their power and exploit the helpless?' He waggled his cigarette. Frank coughed. 'What a terrible world we live in. It's no longer newsworthy for a Cambodian child to lose both legs rescuing their puppy from the tripwire of an old mine. Barely a mention if a wedding party is blown up into red confetti by a drone attack in Afghanistan. But one whiff of chemical weapons and suddenly the world is in uproar. Why?'

'Who cares about the rights and wrongs?' Frank said. 'It's bad business, illegal: end of story.'

'Ah, Frank. I see your difficulty. What if I were to tell you that the main buyers of chemical weapons are not terrorists, but national governments? And in my country, the rules can always be . . .' Pauk bent his spine, bringing his lips to Frank's ear, 'adjusted. Together, you and I could make good business – very, very good business.'

Mad as a fucking hatter. Frank looked around. A statuesque woman walked towards the bar, her blonde wig and low-cut evening dress suggesting she was a regular fixture. He lowered his voice. 'You must be joking. If anyone found out . . .'

'Precisely the problem. Someone already has.'

'Impossible,' Frank hissed. 'Zagrovyl has no involvement—'

Pauk held up a large bony hand. 'Not intentionally, perhaps. By omission.' He finished his drink and shook his head. 'You remember all those materials mysteriously disappearing on their way to Smolensk?'

Only too bloody well. Frank had a bad feeling in the pit of his stomach. Somehow he knew what was coming next.

'I've discovered where they went, what they are being used for.' Pauk signalled to the waiter to bring another round of drinks.

Frank banged down his empty glass. 'Christ-all-fucking-mighty!' He added twice as much of everything to the fresh glass. The ice cubes rattled against the side as his hand trembled. 'There is no way they can link anything back to Zagrovyl.'

'Alas, you're wrong there,' Pauk said. 'You see, your friends have been useful and troublesome in equal measure.'

'What friends?'

'A Dr Jaqueline Silver, for one.'

'No friend of mine.'

'Indeed? She is here, in Kiev. A coincidence?'

'Search me,' Frank said.

'I think it might be more profitable for you to search her. She appears to have acquired something we need. Before it falls into the wrong hands.'

'What the hell are you talking about?'

'What do you know about radioactive tracers?'

'Fuck all,' Frank said, and held up his middle finger. 'And this is how much I care.'

'Really, Frank, I'm shocked. You should pay more attention to the operational details of your own business.'

'I have engineers for the details.'

Pauk shook his head. 'The tragedy of British industry. If the captain no longer understands his ship, how can he be expected to steer it wisely?' He popped a pistachio nut in his mouth, cracking the shell and spitting it back into the tray. 'Well, let me enlighten you.' He wiped his mouth with a paper napkin. 'Tyche, the company you – Zagrovyl – bought, developed expertise in chemical tracking. Designer molecules added in tiny quantities to controlled materials, unique fingerprints lasting as long as the chemical substance does. Harmless to humans, or so they say, but radioactive enough to be picked up by a sensitive tracker.' He cracked another nut. 'And Tyche, that clever company, also makes the spectroscopic trackers, the most sophisticated in the world.'

'So what?' Frank gulped at his drink.

'So successful was the Tyche business model, now all controlled materials must have a tracer added, by law.'

Frank's eyes opened wide.

'You see, Frank, you were right to be concerned about the missing material. You represent the Zagrovyl Group in Europe. You, and you alone, will be held fully responsible.' Pauk smirked. 'Your American masters will not be very understanding. Sack first, ask questions later.'

Had Pauk been listening to every conversation? Stringing him along all this time, waiting to make a point? The bastard. The worst of it was, he was right. If an American soldier was gassed in Iraq and it turned out to be by chemicals from Zagrovyl Europe, there would be hell to pay.

'But fortunately for you, there is one place radioactive tracers fail to work. Where do you think it is?'

Patronising git. Frank resisted the urge to wrap his hands round his long, scrawny neck. Play the game. Get the facts. Assess the risk. 'I don't know,' Frank said evenly between gritted teeth. 'Somewhere highly contaminated.'

'Unfortunately, you rejected the invitation to visit.'

Frank slapped his forehead. 'Chernobyl!'

'You do catch on fast,' Pauk said. 'As I see it, we have a choice. Either we take up this unique business opportunity to work with a rather special organisation. We can manage the risks perfectly well so long as we are in control.'

Frank ground his teeth. What had he done?

'Or, if your appetite for risk is,' The Spider cracked two more pistachios, 'rather more, shall we say, anaemic, then we can cut ties, but in a surgical, precise way, giving both sides a clean break and – in return for a financial consideration – no risk of future association.'

Frank threw back his head and groaned.

'But either way, we need to relieve Dr Silver of the Tyche tracker she recently acquired. Your job, Frank.'

Frank froze. 'Why me?'

'Two reasons.' Pauk emptied his glass. 'First, I need to know I can trust you before I give you your passport back.'

A little spray of pistachio spittle landed on Frank's cheek as he tried to turn away from the vodka-infused stench. He clenched his teeth. The bastard.

'And secondly . . .' Pauk snapped his fingers and the blonde at the bar sashayed towards him. 'I have a dinner date.'

Thursday 2 June, Kiev, Ukraine

Jaq locked the door of her hotel room with a sigh of relief before checking from the window. A steady trickle of traffic, dwarfed by the wide boulevard, moved in both directions as the heavens opened and rain poured down. A few people on the pavement unfurled umbrellas or dashed for shelter, all intent on their own business: no one lingering, no eyes looking up.

She closed the curtains and removed the heavy shoebox from her bag. Under the cardboard lid lay the key to the mystery. She slid her hands between the cardboard and metal and eased the instrument carefully onto the desk.

She'd decided against a new phone, too easy to track, but the laptop she'd purchased in Lisbon came with a set of adapters. She rifled through the padded section of her bag until she found the right connection and plugged the instrument into the mains using her computer lead. Once it was charging, she sat on a chair in front of the desk to inspect it more carefully.

The metal casing was smooth: satin-polished, no sharp corners or rough edges, black with a lilac sheen, cool to the touch. A groove for a USB connector, a protruding switch. Jaq flicked it and the screen lit up. A purple cursor flashed.

Blink.

A prompt? Try return.

Blink.

She ran a finger over the dual English and Russian keypad. What to enter? Try *Sergei*.

Nothing.

Try it in Cyrillic letters. Сергей. No. How about *Test*? No . . . проба? No.

Some problems are best solved by walking away, doing something else, letting the mind wander. Jaq filled a glass with water and sipped it by the window. A few lights shone from the government building next door, but the windows opposite were black. No movement in the darkness. Nothing stirred, just her own shadow.

She switched on the desk lamp. As she sat down the light caught the letters scratched onto the surface of the Tyche tracker. Ключ. Russian for key. Could it be that simple? She typed it in.

Something happening.

That easy? Surely not.

Not.

A new prompt. Пароль. *Parole.* Word. *Password.*

Try *Sergei* again. Сергей. Nothing. Try *Test*? Проба. Nothing doing. Wait a second, what was the number on the locker key? *12016834.* And . . .

Bingo!

Jaq gasped – a map of Europe appeared on the screen. A series of dots formed coloured ribbons stretching between Chornobyl and Western Europe. What sort of map was this? Had someone tracked the movement of materials out of Chornobyl? Had she been wrong about chemical weapons? Was someone smuggling radioactive materials?

She used the touchscreen menu to zoom in. Most dots lay between countries. Did Geiger counters scan trucks passing borders across Europe? Emma would know.

What did the map show? Disposal of nuclear waste from Chornobyl? Or worse, trading fuel for nuclear bomb-making? But Katya said there was no uranium or plutonium left in Chornobyl apart from the material trapped in smouldering Reactor Four. And the really serious stuff was dull to detection, easy to shield in a lead casket, stable until you stuck it in a superheated environment and bombarded it with neutrons. A lump of weapons-grade plutonium was less likely to set off a

Geiger counter than a crystal of cobalt destined for a radiotherapy machine, or a sliver of americium in a household smoke detector, or a truckload of bananas, rich in a naturally occurring radioactive isotope, potassium 40.

But if not plutonium, then what? Low-level radioactive waste sold for dirty bombs? The end destinations didn't denote typical terrorist flashpoints. In fact, they represented centres of industry, but the blobs were big and the countries small, and she could only zoom in so far.

She moved on, swiping across the touchscreen. After the map there were a series of tables, line after line, column upon column of numbers.

The final screens were maps again. One journey per map, by the look of it. This time there was no mistaking the starting point. A faint green spot on the border between Ukraine and Belarus, north of Pripyat. Chornobyl. Then a series of blobs of increasing depth of colour as something moved through Slovakia, the Czech Republic, Germany, the Netherlands, the UK . . . and her heart stopped as she zoomed in. The final bright green blob sat in the centre of Teesside, the heart of UK manufacturing, right at the gate of Zagrovyl.

Was this what Camilla had been investigating? Why the people at Zagrovyl pretended not to know her? Sergei had collected evidence of radioactive material moving from Chornobyl to Teesside through Snow Science. He had requested a meeting; had told Camilla he had the key. And here it was.

What did it all mean? And what the hell was she going to do with it?

Jaq jumped at the knock on the door. With trembling fingers, she moved the Tyche tracker onto the bed and covered it with the quilt. A shuffling noise outside. The squeak of trolley wheels.

'*Domashnye hospodarstvo.*' Housekeeping.

'*Nyet,*' she shouted, no thanks, and held her breath until the noises moved away.

She tossed the quilt aside and flicked back to the map. There were two options. The first was highlighted. She toggled to the second and pressed return. The map changed. It zoomed out and became a world map. The green dots disappeared and now there were dots of many colours. Each ribbon of coloured light connected Chornobyl to a different place: Grozny, Bilbao, Palermo, Belfast, Srinagar, Jaffna, Aleppo, Karachi, Kabul and Pyongyang.

This time the dots were best defined closest to Chornobyl, growing smaller and fainter as they moved further away. The routes were far from direct. They meandered backwards and forwards, making great loops.

Jaq stood and stretched. It was dark outside, the street lights glowing orange, the green and gold onion domes of St Sophia's church lit up from below. The thinning traffic wound its way down the hill, out to the suburbs. People going home. People who had homes to go to. A pain in her stomach, a twisting ache. Was she going to be sick again? She hurried to the bathroom and turned on the tap as she bent over the sink. The pain continued but duller now, not the sudden sharp spasms of food poisoning. Her legs weakened at the memory, and she held onto the sink. Could it be hunger? She had barely eaten for thirty-six hours, hadn't been able to face the prospect. But the idea was suddenly appealing.

Jaq called room service. She ordered clear chicken broth, green tea and plain boiled rice.

While she waited she flicked through the tables on the tracker. There were hundreds of them, but they all followed exactly the same format.

240211 1845 54.597255, -1.201133, 800X, 0C, 800A, 0B
250211 0608 51.126460, 1.327162, 152X, 648C, 152A, 648B,
250211 1145 50.966220, 1.862010, 152X, 648C, 152A, 648B,
250211 1904 48.585741, 7.758399, 152X, 648C, 152A, 648B,
250211 2325 47.799400, 13.043900, 152X, 648C, 152A, 648B,

260211 0606 46.502800, 13.794400, 152X, 648C, 152A, 648B,
010311 0823 46.533200, 15.601100, 72X, 648C, 72A, 648B,
030311 1641 47.45952 18.99284, 72X, 648C, 72A, 648B,
040311 0207 51.532153, 29.575247 Error, Error, 72A, 648B,

The first column had six digits: *240211.* She scanned down; the first two numbers never went beyond thirty-one, the second two up to twelve, so day and month and the last two must be year. The next column was four digits: *1845, 0608* . . . easy – the twenty-four-hour clock format was instantly recognisable. So, the first two columns must be a date and time stamp. The next two columns had seven or eight digits, always six after the decimal point. A familiar format. She let her eyes go out of focus, tried to see the pattern behind the numbers. They snaked and coiled like the ribbons on the map. Maybe, just maybe . . . GPS coordinates? Jaq reached for her laptop, opened a maps application and typed in the last string. On the Belarus border due north of Pripyat. Gotcha!

Were the tables the source data for the maps? Had each GPS location from the table been plotted on the map? It would take time and a fast internet connection to check, but it was certainly possible. Likely, even. But what about the rest of the table? What determined the colour of the dot? Or its size and intensity?

A knock on the door made her jump.

A male voice called out in English. 'Room service.'

That was quick. Her stomach rumbled. Time to eat. She covered the instrument with the quilt. 'Just coming,' she shouted.

Jaq didn't check the spyhole before releasing the chain. She opened the door expecting to see a uniformed waiter with the room service trolley. Instead, a man in a suit stood in the corridor. A familiar figure, but it still took a second to place him. A second in which he inserted a foot between the door and the frame. She gaped at the polished, hand-stitched Italian shoe. Then up, past his pearl-buttoned waistcoat to the pointed chin, the aquiline nose, the cold blue eyes. The first time she'd seen him was in the Zagrovyl headquarters. And with a rush of recognition she remembered the last time, the lift in the hotel lobby here in Kiev.

'Hello, Jaqueline,' Frank Good said.

Thursday 2 June, Kiev, Ukraine

Jaq placed her palms on the door and shoved it with all her might. Frank was too quick; he curled a hand round the outer edge. Principle of levers, force times distance, she had no chance of resisting, even with her shoulder against the door. She jumped back and the door flew open. Frank strode forward.

Jaq placed her hands on her hips. 'Mr Good. Please leave my room.'

He smiled and closed the door, locking it with the chain and standing with his back to it, blocking her exit. 'Not until you give me what I came for.'

'I'm calling security.' Jaq picked up the phone.

Frank lunged forward and ripped the cable from the wall. 'Just give it to me, and there need be no unpleasantness.'

'What do you want?'

'The Tyche tracker. I know you have it.'

'I've absolutely no idea what you're talking about. Get out.'

Frank opened the desk drawer. 'Where is it?'

Hurry up with the soup. 'I am expecting visitors.'

Frank laughed. A short, scornful guffaw. 'I cancelled your room service. There is no one coming, Jaqueline.'

He advanced towards her and she backed away, crossing the room to the window and throwing open the curtains. No escape; the windows didn't open. She banged on the glass, hoping to attract some attention.

'Don't do anything stupid,' Frank said as he continued to ransack the room.

She glanced over to the bed. The quilt was rumpled, the pillows misaligned. She looked away, but it was too late. He followed her gaze and strode to the bed, yanking back the covers to reveal the Tyche tracker.

Frank held the instrument aloft, like a trophy. 'I'll take this. Thank you, Jaqueline.'

'Leave it alone. It doesn't belong to you.'

'Actually, I think you'll find it does.' He leered at her. 'Zagrovyl acquired Tyche and all its assets some time ago. So that would include this tracker. But thanks for retrieving it for me. Personally, I didn't fancy spending any time in a radioactive zone.'

Jaq lunged at him, grabbed the tracker and sprinted towards the door. In the few seconds it took to remove the chain, Frank slammed into her from behind, trapping her between his body and the door.

'Give it to me and no one need get hurt.' His arms reached around her to take the Tyche tracker.

Jaq let go of the machine and shot her hands above her head, leaning back to catch him round the back of the neck. She cocked her head out of the way and slammed his forehead into the door. Once, twice, three times.

Frank cried out with pain as his nose connected with the raised ferrule of the spyhole. She twisted free, dropped to the floor and rolled away, kicking out at him.

He stumbled, but recovered quickly, keeping hold of the tracker. 'You want to play?' He wiped away a trickle of blood dripping from his nose and frowned. 'Don't flatter yourself, you're not my type.'

Jaq backed away until she was at the window.

'Although . . .' He looked her up and down. 'No, sadly, we really don't have time right now.' He laughed. 'Some other time?' He tucked the tracker under his arm. 'I owe you one.'

The door thudded as it closed behind him. Footsteps receded down the corridor. Jaq stood in the narrow gap between the window and the curtain and shivered. The ground was shifting under her feet. On the other side of the curtains, all was not as it seemed. Unstable reality.

The anger rose up from somewhere deep inside her. She threw back the curtains with a roar. Frank's aftershave lingered in the room and she gagged. Clothes lay strewn across the floor where Frank had rifled through her suitcase. The covers from the bed in a pile by the wall, the pillow on a chair.

The overwhelming need to wash drove her, ignoring the mess, straight to the bathroom. She stood under the shower and let cleansing water pour over her. She scrubbed at her naked skin, washing everything away. The water flowed fast, endlessly hot. How did they power the city now the nuclear plant was gone? When the sun didn't shine and the wind didn't blow? Was it gas from Azerbaijan, fuelling the new generation of power plants, or coal from Siberia?

Jaq dried herself and climbed into the other single bed, the one Frank had not touched.

And slept.

Friday 3 June, Kiev, Ukraine

Sunlight filtered through the heavy curtains, moving across the wall until a shaft of light hit the twin beds. One was unmade, chaotic; the other contained a woman curled up in a foetal position, stirring as the light fell on her face. Jaq had slept all night and most of the morning, but now her gnawing hunger could no longer be ignored. She showered and dressed, ignoring the mess. Time to get out and find some food.

She grabbed her bag and stepped out into the street in search of a café, mind blank, one foot in front of the other, one step at a time, stretching her limbs, warm in the sunshine, surrounded by people going about their ordinary business, people who meant her no harm, people who had nothing to do with Zagrovyl or Frank Good. She stopped and clenched her fists, a tremor running though her rigid body from scalp to heels.

Don't look back. Look ahead. Move forward. What next? Avoiding the street sellers in the park, nauseous at the thought of watermelon, she advanced towards an avenue of trees, walking briskly until she reached the Lavra Monastery, green and gold and white, sparkling in the sunshine. Last night's rain had washed the city clean.

At a tourist café she ordered black tea and dry toast. Sitting at an outside table, under the shade of a horse chestnut tree, Jaq observed the comings and goings, focusing on minutiae to keep dark thoughts at bay. A steady stream of tour groups – some on foot, some in buses, some with guidebooks and some with guides. The toast arrived; it crunched and exploded into insubstantial crumbs in her mouth. She chewed slowly. One step at a time.

Above the pale green leaves, flower candles were blossoming: pearl-white buttons. A sneering face formed in the canopy, the palmate leaflets pointing at her like mocking fingers. Where was Frank Good now? Had he left the country? Or was he still lurking somewhere in Kiev? Still at the hotel? Should she confront him? Report him to the police? And what would she say? *Frank Good assaulted me and stole something. And why did you wait until today to report it? I was tired. And how did you come by the object he stole? A prostitute in Minsk gave me a key to be used in the Chornobyl zone of alienation, and I found the tracker belonging to a Ukrainian explosives expert who went missing in Slovenia where I am wanted on murder charges.*

Yeah, right.

It didn't take long to locate the person she had come to find. Petr, the first guide from the Chornobyl tour, leading a group out through the gates of the monastery. She waited until he had ushered them onto their departing coach before standing up and waving.

'Jaq!' Petr rushed over, grinning from ear to ear.

His delight was touching, but she clamped down on her own furtive fizz of pleasure. Strictly business. She indicated the chair opposite. 'I believe I owe you lunch.'

'A bit early?' He checked his watch and then smiled. 'But why not? Look, I know a better place, a short walk. You can tell me what happened on the way.'

She paid for breakfast. They strolled side by side under an avenue of trees, the heady blossom wafting through the air, a sensual scent of new life, new beginnings.

Petr listened as they descended the hill, never interrupting. She chose her words carefully, struggling to limit her story to the bare bones of watermelon sickness and overnight recovery. When she fell silent he asked her about herself. Evasion became harder, and she was grateful when they turned into a narrow alley and had to continue in single file.

The restaurant, hidden between tall buildings, announced itself with mouthwatering smells. The owner, a rotund man with an impressive, curling moustache, greeted Petr like an old friend, clapping him on the shoulder and beaming at Jaq as he hurried them to the terrace. The outside tables, tucked under a wooden trellis heavy with vines, commanded a splendid view of the River Dnieper.

'There's something you aren't telling me.' Petr waited until the owner had bustled away. 'What were you doing with that Geiger counter?' He leant forward and made eye contact. 'I honestly don't remember you bringing one in.'

'It's a long story.' She avoided his gaze. 'Thanks for recovering it for me.' No point mentioning that Frank had stolen it from her.

'The least I could do after abandoning you in Pripyat.' He gazed out over the river. 'How are you feeling?' He nodded at the menu. 'Recovered enough to try one of Igor's specials?'

'I've barely eaten since the watermelon that caused all the trouble.' She patted her stomach. 'And now I'm starving!'

A waiter pirouetted past with a bowl of steaming stew, delivering it to a man in dark glasses. The smell of pork and cabbage lingered. Her mouth watered.

Petr followed her eyes and laughed. '*Solyanka.* Want to try some?' He signalled to the waiter.

When the stew arrived, she couldn't get the spoon into her mouth fast enough. Cubes of seasoned, tender pork in an unctuous liquid thickened with potato. A potpourri of onions, carrots, swede and dense fresh and sour cabbage added at the last minute to keep shape and texture and a slight crunch. If it hadn't been piping hot, she'd have gobbled it down faster. As it was she had to blow on each forkful. The more she ate, the hungrier she became. When she'd finished, she wiped the bowl clean with black bread. The waiter offered her more, but she shook her head and sat back.

'So, Jaq, where are you from?'

The passport she had used to book the Chornobyl tour was Portuguese. Best to stick with that.

'I flew from Lisbon.' Via Slovenia, Teesside and Belarus.

'You don't have a Portuguese accent.'

'I studied and worked in England.' Before Slovenia.

'England, eh?' Petr accepted a refill of stew. 'That woman was talking rubbish.'

Which woman? Surely not Katya, the guide. She stared at him for a clue.

'It's true, *chornobyl* is the name of a plant,' he waved a fork in the air, 'but it isn't wormwood.'

Ah, Megan, he was talking about mad English Megan.

'It's mugwort,' Petr continued. 'Some Russian started the rumour because he confused the English botanical names. Wormwood is *polyn hirky* – bitter artemis or *artemisia absinthium.* It's what the French use to flavour absinthe. Mugwort is *chornobyl* or *artemisia vulgaris.* There are no biblical predictions about mugwort.'

'Are you a botanist?'

'Mycologist.' He speared a cube of pork and held it aloft. 'We're studying fungi tolerant to radiation. Many new species thrive on it. They grow in the dark inside the reactor sarcophagus, using the energy from radioactive particle decay instead of the photons from light.'

'Nature finds a way.' Jaq leant to one side to let the waiter remove her empty bowl, shaking her head at the offer of dessert.

'It's an important discovery.' Petr put his fork down and patted his belly, signalling to the waiter that he was also finished. 'Could be a way to feed space travellers.'

'What – atomic-powered spaceships with mushroom farms feeding on nuclear waste?'

'Something along those lines. How about you? What did you study in England?'

'Engineering.' That usually stopped the conversation dead.

'Nuclear?'

'Chemical.' She folded her napkin and steepled her fingers.

'So, what are you working on?'

'I'm between jobs at the moment.' Sacked from one, suspended from another. Time to deflect him from this line of enquiry. 'So how come you moonlight as a tour guide?'

'I love my work, but . . .' Petr rubbed his forefinger against his thumb, '. . . it doesn't pay much.'

'Do many people actually work inside the zone?'

'Thousands,' he said. 'All the construction workers on the New Safe Confinement, hundreds of them, in shifts. Three weeks on, three weeks off. Then there are the epidemiologists, zoologists, botanists and mycologists. People like myself who come and go.'

'Does anyone live there permanently?'

'Not officially.' Petr looked over his shoulder. 'But, yes. Some old folk returned to their farms. They were told of the danger, but I guess they decided a slightly increased risk of cancer was better than the absolute certainty of grinding urban poverty. We don't really understand why radiation affects people differently. But the younger you are, the worse it is. The old peasants who went back are the survivors. If they've lived this long, there's every chance they'll live longer.'

'So only old people?'

Petr shuffled in his seat. 'There's been a change. People fleeing war in Azerbaijan, Georgia. I guess the zone looks like a paradise compared to where they came from.'

'But how do they get in?'

'The security is pretty tight on the Ukrainian side, as you saw, but it's more porous in Belarus. In fact, I collect samples there – they're so used to my bike, they never stop me.'

'Motorbike?'

'I have an old Ural. You drive?'

'You bet!' She paused. 'Ever been on the border due north of Pripyat?'

Petr pulled his seat closer to the table. 'Why do you ask?'

'Curiosity.' She twirled a strand of hair round her finger.

'No, why that exact location? Come on, what else are you not telling me?'

What to say? Was Petr a Zagrovyl stooge as well? He was certainly good-looking, intelligent, unusually interested in her. Another man for hire? Careful.

'Have you been there?' she asked.

'This is extraordinary. You have identified the one area I've never been able to take samples from. Last time I tried, I had a gun pointed at me. Hunters protecting their patch. It's illegal to hunt in the zone, but there are rich men who will pay and poorly paid bureaucrats who will look the other way for a price. They come in by helicopter.'

A tingle of excitement. 'Did you ever come across a helicopter pilot, a guy called Sergei Koval?'

Petr shook his head. 'I've only been working here for a couple of years.'

'He isolated that location as an area of . . . special interest.'

'So, this isn't a tourist visit?'

'Business and pleasure,' Jaq said, and smiled.

Petr grinned back, and then dropped his eyes and folded his hands. 'Fancy a spin on the bike?'

Christ, this was so easy. Too easy? Another trap? What the hell. 'Yes.'

'Where do you want to go?'

'Guess.'

Petr laughed. 'Back into the zone?'

Friday 3 June, Kiev, Ukraine

'Want to drive?' Petr handed her the keys.

Jaq zipped up the borrowed leather jacket, tightened the chin strap on the helmet, pulled the visor down and straddled the bike. Once Petr hopped on the back she kicked away the stand, clunked into gear and opened the throttle.

It had been a while. The Moto Guzzi 500 cc bike had languished in the garage for much of her marriage. Gregor was a terrible pillion passenger, giving unwanted advice, trimming the bike on corners. She sold the motorbike to keep the peace. If only she'd banished Gregor instead.

They sped past faceless concrete suburbs on wide boulevards and joined a minor road that wound through gentle hills and ancient forests. Jaq put the bike through its paces, tilting into the curves – one knee almost on the tarmac; accelerating out of bends – steering with the throttle and harnessing the raw power of old-fashioned Soviet engineering. So good to be on the open road again with the sun on her back, breeze in her hair, exhilarating speed, thrilling control and a steady vibration between her thighs. A sonata of speed: the whistling wind, the deep bass of the engine roar, squeals of friction, rat-a-tat-tat of gravel and stones spraying out behind her. A whiff of burning rubber and unburnt fuel mixed with the woody terpenes and sweet esters of the forest. She barely noticed Petr behind her. He shadowed her every movement, holding onto the luggage rail at the back with one hand, keeping his body at a respectful distance.

The traffic thinned as they headed north. Jaq patted the pocket of the borrowed jacket, her passport with the Belarus visa still there.

Over the border, the sun lit up the countryside. The biggest threat to the planet was human. Left alone, the plant and animal life sprang back with vigour. Even new fungi evolved, feeding on the radioactive mess humankind left behind.

As they approached their destination, she weaved around potholes filled with vegetation, slowing down as the road surface disintegrated. A doe and her fawn skittered across the forecourt of an old petrol station. Unmanned and deserted but in surprisingly good condition, gleaming pipes and quick-release couplings beside a large area of hardstanding free of weeds. Struck by the incongruity, she made a mental note to ask Petr what it was used for. Soon after, the road became an earth track and faded away entirely, overrun by bushes. She brought the bike to a halt.

Unprepared for the brutal energy of nature left to its own devices, Jaq gazed ahead in wonder. So, this was what happened with unfettered competition – anarchy. Without a managed division between forest and farmland, wetland and dry, the fastest-growing plants invaded. A tangle of willows and reeds blurred the edges of lakes, thorny bushes choked saplings and creepers sucked the life from mature trees. All that remained was a chaotic tangle of stunted and straggling plants where once there had been the order and beauty of productive farmland and forest.

Petr jumped off and helped her rock the bike onto the centre stand. 'On foot from here,' he said, removing his helmet and jacket. 'It's quite hard going. Leave everything you don't need with the bike.' He recovered her bag from the side pannier and handed her a portable dosimeter, clipping another onto his jeans.

Jaq peeled off the leather jacket and luxuriated in the warmth of the sun on her skin. She removed her passport and zipped it into the front pocket of her Tardis bag and converted it into a backpack.

Petr stuffed the helmets and jackets into a dry bag which he sealed and strapped to the bike. 'It's hot, but tuck your jeans into your socks and roll down your sleeves.' He handed her a bottle of water. 'Try not to touch anything. There are a couple of radiation hotspots here,' he warned. 'Everything contaminated should have been buried, but some people hid stuff to try and take it out later. I've set the meter to warn at ten, okay?'

There was something reassuring about Petr. The fact he'd let her drive, the way the guards waved him through the border checkpoint. His story seemed to map out.

'This way.' Petr nodded at an opening in the trees where the old road snaked out of sight. 'Let's go.'

Friday 3 June, Chornobyl Exclusion Zone, Ukraine

A gentle breeze swished through the forest in the zone of alienation. Petr led the way as the road petered out and Jaq followed him through the thickening trees. He stopped to collect a sample of bracket fungi from the rotting bark of a fallen tree and she continued on until she reached a high wall, so well hidden in the trees she almost bumped into it. Expecting it to be a ruin, she followed it to the left, but it extended as far as the eye could see. She tried to the right, but it was huge.

'Petr,' Jaq shouted over her shoulder. 'What's this wall?'

Petr came crashing through the undergrowth. He sucked in his cheeks and made a popping sound of surprise. 'I think there was a factory here once. It was abandoned long ago.'

'I'm going to explore, see if there's a way in.'

'There's nothing in there.' He pointed back down the path. 'I'm going to take some soil samples. I'll find you when I'm done. Keep near the wall. Don't go too far.'

Jaq checked her portable dosimeter. Tolerable. About the level of radioactivity given off by granite. No worse than strolling down Union Street in Aberdeen. Considerably quieter, but harder going over peatbog and scrub. She shouldered her way past a stand of young birches, the branches bending and swishing, leaping from ankle-twisting, squelching hummocks of moss to hillocks of bramble and blackthorn that tangled and tripped. The first entrance she found in the wall was bricked up. A long time ago, judging by the wild honeysuckle clambering all over it. She tried to track what was left of the road leading from it, but it was overgrown with prickly shrubs: cranberry and bearberry.

Kweee-kwee. A little bird, the size of a thrush, whistled an alarm from the top of a larch tree. Another bird, pearly grey with a white breast and black mask, hopped along the top of the wall, five metres above the ground. *Chek-chek-chek.* Keep away. She ignored the warning and continued her trek to the next corner. *Trr-trr-trr.* The rattling was insistent, ending in an explosive *aak-aak-aak* as the little bird flew away.

She wiped the sweat from her brow and paused to retrieve the water bottle from her bag. After she turned the second corner, a hush descended, strangely silent: no chirp of crickets, no more birdsong, no rustle of small animals running away from big feet ripping through the scrub. Perhaps the wildlife had the sense to hide from the sticky heat of the afternoon. But plants as well? The ivy on this side of the wall straggled, yellow and sickly. From the scorched and blackened earth rose a stench of garlic and putrefaction. A forest fire, perhaps, though it smelt of rotten fish rather than burnt wood. It made her route easier, through dead trees and bare shrubs and snap-dry stalks. She hurried past another bricked-up doorway and ploughed on.

A honeyed scent hit her as she rounded the third corner; she took a deep breath. This side of the impregnable fortress was a riot of coloured leaves and flowers struggling though the brambles and birches. Had there been a garden here once? What was left of the delicate ornamental plants – a blood-red acer, a pink blossoming azalea, a yellow rose – were being strangled and choked by the native species. Without a gardener to weed and cut and select, only the strongest and most aggressive survived. There was little hope for the delicate and beautiful.

She sucked water through a closed straw – just a mouthful. Flies buzzed around her head. She swatted them away and completed the circuit, returning to her starting point.

Jaq shivered. There was something eerie about the old factory. The wall stretched for about two kilometres, five hundred paces on each side of a square, yet it was in remarkably good shape for an abandoned structure in the middle of a forest; some of the mortar looked as good as new. An entrance and an exit, both long bricked up. No way in or out. And yet something felt wrong. Smelt wrong. Smelt fishy.

She walked away from the trees, looking for Petr. The beats in the air started softly and grew in intensity. The swishing saplings swayed and bent. Helicopter approaching, *merda!* Petr had warned her about hunters. She crouched down as the first one appeared, a black metal insect – whirling blades chopping the air – banking as it flew over the wall. A second one followed, circling wider, coming right overhead. Her heart beat faster. Would they spot her? She didn't want to be mistaken for a deer. Move back to the wall? Better to stay still. She dropped to the ground. Ouch! The bramble thorns were razor sharp, piercing her palm. Stupid, she should've worn gloves. She licked her hand, pulling out the thorns with her teeth, gagging on the taste of salt and rust as she spat out the blood. The cut hurt like hell, but she focused on the noise directly above her head. How many helicopters?

Whack! Jaq jumped at the sharp noise. What was that? Someone felling a tree up ahead? With an axe? *Whip-crack!* Whistling in her ear. Insects? No, too high-pitched and moving too fast for beating wings. A sapling exploded beside her with a deafening crash. *Merda.* She flattened herself, face down on the ground, elbows tight against her side, clammy hands clutching trembling legs, her insides liquid again.

Bullets. They were shooting from the air. Shooting at her. Bad. Very bad. The trail through the undergrowth would be visible from the broken twigs and branches, the trampled moss and shrubs. The shooters must think they were following the track made by a wild animal. And where the trail stopped it would signal her current position. Quick. Move. She slithered forward using her elbows and knees, moving away from the wall at an angle to her most recent trail. A cluster of willows lay ahead, peeling white bark and a pale green canopy. The young trees would only come up to her shoulders when standing but would provide better cover.

Her hand was bleeding, and thorns and twigs scratched at her face and arms as she hauled herself forward. Even hotter so close to the ground, airless.

Inside the cave of willow, she made herself as small as possible, crouched in a foetal position over her bag, not daring to look up. The downdraught from the helicopter moved the leaves as it hovered, the little trees rustling. Could they see her hiding place from the air? The helicopter hovered close now; she could hear voices shouting above the chopping blades.

Her logical brain told her to remain as still as possible. She clenched her jaw and tried not to jump at every crack and blast and thump and boom. She so desperately wanted to flee, stand up and run from the deafening, terrifying noise all around her. Run where? Straight into the line of fire? It took all her strength to stay still.

Was this what it was like to be hunted? Did animals feel this dread, this desperation? Did their breath catch in their throats, choking and gasping against the fear? Did their hearts pound in their chests, rat-a-tat-tat, like the bullets all around her? Did their insides turn to water? All to provide men with entertainment.

But were they hunters? Or worse, was someone determined to hide what went on inside that wall? Hunters or security, any movement would attract more bullets. The only option was to stay perfectly still. Jaq closed her eyes and tried to slow her breathing. Focus on something else. She looked around.

It took her a moment to realise what the movement was. Just above her head. A wriggling animal. A field mouse. Impaled by its neck on a thorny branch. Bleeding. One last spasm and then still. She looked away only to see the corpse of an amphibian. Partly skinned. Those little birds must have been shrikes, the sentinel butchers. And this was their larder. Nature red in tooth and claw. What could be more natural? Natural doesn't mean gentle or kind. It just means survival. Kill or be killed. *Chiça*, she needed air, needed to get out of here.

The whir of blades receded. She crawled forward, further away from the wall when she sensed the new danger. The hairs stood up on the back of her neck and a trickle of sweat ran down her temple. She stopped moving and listened. No mistaking it, the sound of boots squelching over wetland: men on foot. Where had the helicopter landed? *Thud, thud*, the earth beneath her was vibrating. Footsteps getting closer. If they were hunting wild animals, it was bad enough. If they were protecting a location, it was even worse. What had she been thinking, to come here? She lay flat, a little mound of sphagnum pillowing her face, the groundwater seeping into her clothes, slowly sinking.

Would anyone notice if she didn't return? She had no family left who cared. Her mother refused to acknowledge her daughter, still crying for her long-lost son. Emma and Johan might become concerned, but they were used to her long absences.

The boots were getting closer. She sank further into the peatbog and held her breath.

Friday 3 June, Chornobyl Exclusion Zone, Ukraine

A noise behind her. Something slithering through the undergrowth. Jaq turned in alarm then sighed with relief. Petr – *graças a Deus!*

'Stay where you are,' he whispered. 'Whatever happens, stay here until I come for you. I heard them talking. Russian hunters. Let me handle this.'

He slithered forward. '*Ne strelyach!*' he shouted. *Don't shoot!* He raised his hands above the grass and let his body follow as he emerged from the undergrowth. He moved away from her, towards the hunters.

The conversation was short. She could hear Petr talking calmly in Russian. She imagined him showing the hunters his pass, the bagged samples of interesting fungi, reassuring them of his scientific credentials.

'Where is the other one?' English. Heavily accented. Spanish? Latin American? 'There were two helmets on the bike.'

'He says he came alone.' The Russian spoke good English.

Jaq prepared to stand up. She couldn't let Petr face this unaided, couldn't oblige him to lie on her behalf.

'Then he's come for the Englishman,' the Latino said.

The Englishman? Jaq remained where she was.

'I don't think so, Mario.' The Russian spoke calmly. 'I've seen this guy before. He's definitely one of the Chernobyl scientists.'

'I don't care.'

Jaq was on her knees, her hands on the grass, ready to push herself into view when the shot rang out. She collapsed back onto her stomach, a primeval form of self-preservation taking over. Quick, shallow breaths. *Ó meu Deus*! Had they just done what she thought they had? There was no sound from Petr. Maybe it was a warning shot. *Whoomph* – an exhalation without sentience – a last breath and slow collapse. Jaq recoiled, shrinking into the smallest, tightest ball, hands over her ears, unable to deny the thudding vibration of a body hitting the ground.

'Christ, Mario.' The Russian sounded more peeved than shocked. 'Was that really necessary?'

'He saw the complex. He saw The Spider.' The thump of a boot kicking something soft.

Jaq cringed. She swallowed back the bile in her throat, fighting the urge to retch. Keep quiet.

'What now?' Another voice, Russian.

Jaq backed away, slithering into the marsh, retreating inch by inch, freezing at the rasp of a match striking a box.

'Take the bullet out. Crash the bike and make it look like an accident.' Mario paused and puffed repeatedly. 'Smash him up good and proper. Then set him and the bike on fire.'

The bastard. The cold, wicked, murderous bastard. The stomach spasms became stronger as the scent of dark tobacco filled her nostrils. How dare he smoke a cigar beside Petr's corpse? Quiet! Keep quiet. Jaq put her hand to her mouth and gagged it with her fist. If they hear anything, you're dead.

Jaq cowered in the marsh, her heart hammering, her limbs trembling as barked instructions led to retreating footsteps, the rattle and whoosh of some machine, a creaking, grinding noise, a metallic thud.

And then silence.

Friday 3 June, Chornobyl Exclusion Zone, Ukraine

Peaty water seeped over her body, the soft ground giving way as she sank. But Jaq dared not move. The motorbike engine spluttered from the road as someone repeatedly tried to kick-start it. Not Petr. He wouldn't flood the engine.

Petr was dead.

Cabrões. Someone was going to pay for this.

As the shock ebbed, a swell of unreal calm flooded over her. A staged motorbike accident. Smash him up. Set fire to him. The keys to the bike were in his pocket. The bike was going off the road along with Petr.

No bike. They'd travelled perhaps twenty miles since the last habitation. She'd be visible on foot, an easy target on the road. In any case, she didn't have enough fresh water left for a twenty-mile hike. *Water, water everywhere but not a drop to drink.* She adjusted position and sank further. She'd have to wait until dark. What new dangers in the dark? Lynx? Bears? Wolves?

Why would anyone shoot Petr? A harmless mycologist, a threat to no one. *He's come for the Englishman.* What Englishman? *He's seen The Spider.* Who or what was The Spider? *He's seen the complex.* What complex?

She looked back at the wall. It might be a trick of the light, but further away from the road, the signs of recent repair were more obvious. This was no ruin. Whatever it was enclosing – hiding – was worth killing for.

Where had the men gone? The footsteps moved away from the wall, not towards it. There had been no new whir of a helicopter overhead, so the men couldn't have been hoisted up off the ground. No sound of vehicles either, and the road was not far away. Was it a trick? A trap? Were they waiting for her? Jaq's breath came in hot bursts, bubbling from her mouth into the marsh. Her neck ached as she struggled to keep her nose above the waterline. The bog wrapped its cool arms around her, claiming her, engulfing her in thick brown water. If she didn't move soon, she would drown.

Squelch. She dragged one hand out of the bog, *slurp*, then the other, and inched her way out of the peaty water, back to the hummocks of moss. She ran a wet hand through her hair; it caught on sticky seeds and sharp burrs. She listened. Nothing but the buzz of flies and whine of mosquitoes. Eating her alive. When she could bear it no longer, she straightened up and peeked above the grass. No one there; the men had gone. Her whole body shook with relief.

A beech tree towered over the surrounding saplings. Smooth bark and low branches. A possible refuge until nightfall. Jaq shimmied up. Hard work to reach the first branch, then easy to climb. She surveyed the area from her new vantage point. *Credo!* So that was the complex.

Inside the wall were several low buildings, invisible from outside. The roof looked new, no sign of moss or broken tiles. She climbed higher and examined the walls. Painted aluminium cladding, double-glazed glass windows, more 2010 than 1980s. But what was it for? She'd almost completed a tour of the perimeter wall before the helicopters arrived and still had found no sign of an entrance or a supply road. How could the complex be in use if there was no road in or out?

Freeze. A noise. Voices. From behind the wall. She climbed as high as she dared and craned her neck, trying to locate the source. Definitely coming from inside the complex now, and it sounded like Mario, the bastard who murdered Petr. And then other voices.

'Let me go!'

Jaq gripped the branch. There was no mistaking Frank Good, no confounding his imperious English accent. So, he had been in league with the murderers and now it had turned sour. A happy ending.

A female voice, too, the sing-song cadence and soft timbre vaguely familiar, but Jaq couldn't make out the words.

She remained in the tree as the shadows lengthened. Her clothes dried quickly, but the stench of the bog attracted flies. They were everywhere, swarms of tiny grey flies, fat black horseflies, all buzzing and whining and stinging. *Caramba.* How long could she stand it? *Valha-me Deus.* What choice did she have?

When it was dark she would run for help. The shortest way to habitation was south-east. It would take her deeper into the zone of alienation, but there was more cover and she could use the deserted nuclear power plant structures as waypoints to find the dormitory in Sector Twelve. How far away? Five, ten miles? She couldn't risk the forest road north into Belarus. Dehydration and wild animals posed a far greater risk than radiation.

She slithered back down the tree, then paused. The rattling noise followed by grinding metal. A pause. Then a metallic thud. Footsteps. New voices. Close, this time. She straddled one of the low branches and froze.

'You are sure there was no one else?'

A new voice. Deceptively soft, but there was no mistaking the authority. He spoke in English with trilling, hissed sibilants and a strong Russian accent.

'Positive.'

Jaq recognised Mario's voice, the cigar-smoking Latino murderer.

'And you've cleared up?'

'No trace left, sir.'

A note of deference from Mario. The hissing Russian was the one who called the shots. Checking for himself. She pressed herself against the smooth branch as the swish of the grass became louder. Don't look up. Please don't look up. She held her breath and peered through the leaves. The hissing Russian stood directly underneath, right below the beech tree, so tall his dark spiky hair, cut short around a bald crown, almost brushed the lower branches.

He held his head at an awkward angle to his shoulders, his limbs long and thin, moving sideways in short, jerky movements. So, this was The Spider. A perfect nickname.

'We're leaving,' he said. 'Anton and Victor can stay behind.'

'And our visitors?' Mario asked. A match flared, and cigarette smoke swirled in her nostrils. Perhaps they only broke out the cigars after murdering someone. A mosquito whined in her ear. A high-pitched warning. She daren't move a hand to swat it.

Jaq slowly let out her breath as the two men moved away.

'The Englishman could be useful,' the hissing Russian said. 'On the plus side, he's from Zagrovyl. Unfortunately he's a stuck-up, inflexible prick.'

Too right. They had the measure of Frank Good.

'He's resisting.'

Resisting what?

'And if he won't play ball?'

So, there was some doubt? In which case, why had Frank stolen the tracker from her?

'He stays here,' the spidery Russian said. 'Interrogate him any way you want. Then kill him.'

Saves me a job.

'And our Swedish expert?'

Jaq froze.

'Dr Camilla Hatton travels with me.'

Friday 3 June, Chornobyl Exclusion Zone, Ukraine

The leaves of the beech tree shivered. Twigs snapped, the branches suddenly too narrow, the foliage suddenly too sparse to hide the woman falling from the canopy.

The shock slammed into Jaq, gripping and twisting her stomach. Heart swelling, bursting from her ribs, bile rising to her throat choking her as she struggled to catch her breath, an ache throbbing at the back of her throat at each rasping breath. And then the shivering began. She felt herself slipping, falling.

The Swedish expert.

Camilla was alive.

Camilla had lied from the beginning. Camilla had threatened her, interfered with her work. Yet all along Jaq had given her the benefit of the doubt, found an explanation, an excuse. Convinced herself that Camilla had perished trying to destroy the bad stuff.

She could fool herself no longer. Camilla had masterminded the supply of chemicals from Zagrovyl. She was in league with the murderers inside the complex.

Murderers.

Murderers who would put a bullet through her head, just like they did with Petr. Jaq hooked a narrow branch with her knees just in time to break her fall, swung up and stretched out full length, peering down. Had they heard her?

Breathe. Breathe.

The men were gone. The rattling, grating, grinding of metal on metal was followed by silence. The minutes passed, then the same metallic sound came from the other side of the wall. Voices as Mario and his Russian boss emerged inside the complex. A tunnel? Under the wall. The only possible way in.

Jaq climbed back up into the tree, but she couldn't get high enough to see in.

'*Vamos, poydemt!*' Let's go!

At the whir of accelerating blades, Jaq crouched low among the leaves. Would they see her? She lay flat against the smooth, wide branch and held her breath, letting it out as a helicopter rose from the complex and flew away. Only one. The other chopper was still inside.

Time to go. Time to get as far away from this cursed place as she could. Time to report Petr's death. Time to get the authorities involved.

The shout rang out from behind the wall. 'You tricked me!' Then a scream. Frank Good, no doubt about it. She loathed the man with every sinew in her body, every fibre of her being. Let the bad guys sort him out.

If she'd been so wrong about Camilla, what else had she been wrong about? Could she have misjudged Frank as well? Could she really leave him there, knowing he would die?

Yes, easily. He sent Beige to spy on her. To kill her. Ben had almost drowned. Frank Good deserved everything that was coming to him.

The sooner she slipped away, the sooner she could bring in backup. She'd go to the British embassy in Kiev and tell them everything. What about corroboration? Petr was dead. Oh, Petr! The only person who could back up her story was Elena in Belarus. Not the most credible witness, hardly someone you would rely on in a tight spot. The one person they might listen to was Frank Good, but he was about to be tortured inside a deserted complex in the zone of compulsory resettlement near Pripyat with no road in or out.

If only she had some evidence.

Jaq dropped from the tree onto the soft moss with barely a sound. It wasn't hard to see where the men had gone. A gold and black cigarette end lay smouldering on the grass. Black Sobranie.

The spidery Russian had expensive tastes. She kept as low as her trembling muscles would allow and followed the broken stalks and bent grass until the trail stopped beside a clump of trees. Where was the entrance to the tunnel? A dead tree, larger than the saplings, stood out. It had been struck by lightning, split down the middle with a charred centre and hollow branches. She inspected it, running her hands over the rough bark. When her fingertips detected a smooth patch, she investigated more closely. The button was at shoulder height, concealed behind a hinged flap of wood. She glanced around and made sure she could find it again later, when she returned with help.

Her fingers searched for the dosimeter. Cumulative dose 85. She shook her head. She needed to distance herself right now. Clean up, destroy her filthy, contaminated clothes, find a cold drink and some food. She started to creep away.

'No!' A scream. 'What do you want from me?'

Jaq stopped. What they wanted from Frank was information. While she went for help, he'd spill the beans. He would sing. Long and loud. The Russians had let her slip through their fingers in Kranjskabel. But Frank would tell them Dr Jaqueline Silver was not locked up in Slovenia, had found the Tyche tracker, had looked at the maps, had seen the evidence of what they were up to. They wouldn't let her escape a second time. They would come for her as well. If not today, then next week; if not next week then next month, if not openly in the day, then by stealth at night. She knew too much. She would never be safe again. Unless she stopped them now.

Stop them how, exactly? Unarmed against brutes with guns and helicopters inside a fortress. How many? The Russian boss said he was taking everyone. Leaving only Anton and Victor. One against two. Would Frank be any help? If his life depended on it? Two against two. Another shout of pain tore through the air like a thunderclap.

Saving him was one thing. Saving herself was another.

Who was more likely to be believed? A woman accused of professional incompetence, manslaughter and murder? A woman who had skipped bail in Europe to come east? Or the operations director of a FTSE 100 company?

Face it: if she was to have a future, she had to get Frank out of there.

She returned to the clearing, reached out for the panel in the tree and pushed the button.

Friday 3 June, Chornobyl Exclusion Zone, Ukraine

This time, Jaq was expecting the noise: first the rattling and swishing, then a metallic grinding and squealing. Exactly the noises that preceded the men disappearing. She jumped aside as the ground vibrated and bulged and a square began to pivot away from the forest floor. The trapdoor opened in juddering, creaking lurches, revealing a metal underside with a cap of soil and grass clinging on tenaciously as it rose to an angle of 45 degrees.

Jaq peered into the abyss.

Light reflected from an iron ladder, the steps shiny and worn with use, stretching from the lip of the opening to the concrete floor below. Beyond it a tunnel snaked towards the complex. She hesitated. How long did she have before the trapdoor closed again? Had she already activated some sort of alarm inside the complex by opening it? Too late to worry now. She clambered in.

Inside the tunnel it took a moment for her eyes to adjust after the bright sunshine outside. It wasn't dark; a ribbon of dim lighting ran along a square ventilation duct.

The tunnel was large enough for her to stand upright. She ignored the cobwebs brushing her face. After a few hundred metres, the tunnel branched into six. She tried to remember the layout above ground and chose the direction most likely to continue into the complex. It began to slope upwards, and she moved more slowly as the height fell. She was almost bent double when she came to the first dead end.

Jaq turned back and entered the next most likely route, to the left of her first choice. The second tunnel remained broad and high, but it sloped steeply downwards and turned a sharp left. She was almost sure this would take her away from the complex, so she retraced her steps.

When she arrived at the starting point, she had to think carefully to remember which tunnel she'd come from and which to try next. Something more systematic was required. She found the penknife in her bag, flipped it open and scored two marks in the earth wall at the entrance to the tunnel she'd rejected, then selected a third tunnel, the second on the right. Before entering she made three scores above the first lamp.

The third tunnel ran straight for several hundred metres. Her internal compass told her it was the right direction, should bring her under the wall and into the complex. At the end was a ladder. She climbed up and knocked her head against a metal trapdoor. However hard she pushed, it wouldn't budge. Where was the mechanism? Her hair brushed against a piston cylinder, similar to the one at the entrance. Where was the button to operate it? She searched for a control panel.

Time to retrace her steps. As she hurried back along the tunnel, sweat trickled from her temples. She wiped her brow. Something had changed. The cool breeze had stopped. The ventilation ducting was no longer rattling. A grinding noise took its place. *Merda.* She ran towards the noise, reaching the ladder as the trapdoor slammed shut.

Not good. Where was the internal control for the trapdoors? *Meu Deus.* Why hadn't she checked first? Jammed a branch in the opening before she descended into the labyrinth? What if the trapdoors could only be operated from the outside?

She went back to each of them in turn. There must be a way. Surely there was a way. There was no breeze in the tunnel, no natural light; the whoosh and rattle of forced ventilation had long ceased. Within minutes the stuffy underground warren smelt of decay. And death.

The lights went off. Darkness closed in.

Friday 3 June, Chornobyl Exclusion Zone, Ukraine

Jaq had a healthy respect for confined spaces. She didn't suffer from claustrophobia, and she was no more frightened of being trapped in a dark underground tunnel than most people, perhaps less than most as she'd been underwater caving in her time. But from her working life she recognised the danger.

The story usually goes like this. Man climbs into a confined space. Collapses. His mate shouts for help, peers in, can see no obvious danger, goes in after him. Collapses. The rescuers arrive. They guess the oxygen is low. Reckon they can hold their breath long enough to get in and out, quick as possible, save their mates. They can't. They collapse too. One rescuer after the other. Multiple fatalities. All from lack of oxygen. The air we breathe is about 21 per cent oxygen, 78 per cent nitrogen and 1 per cent argon and other stuff. At 17 per cent oxygen, you get a headache. At 15 per cent your heart races but you slow down. At 10 per cent you lose consciousness and are dead within an hour. Below 6 per cent, you die within a minute.

Jaq recognised the signs of hypoxia. Pulse racing, head thick and stupid, breathless and exhausted. She sucked in great gulps of air, but nothing satisfied her gasping. She'd probably been down here for three hours. When she first did the calculations, she wasn't too alarmed.

She sat down and remained still to conserve oxygen. She'd been to the end of every tunnel, glad of the score marks that stopped her repeating the same exercise. Each of them ended in a ladder and trapdoor. Each of the trapdoors was tight shut. Including the one that had allowed her to enter this labyrinth.

She followed the ventilation ducting, pulling off the grilles one by one, looking for an alternative way out. The duct diameter was too small to enter and without the fans, marshy gas was seeping into the tunnel.

The dosimeter on her lapel started beeping. Alarm point. Great: if the lack of oxygen didn't kill her, maybe the radiation would.

How had she got herself into this situation? She'd come close to death before: kayaking rivers in flood, skiing over cliffs, paragliding, driving a motorbike, handling explosives. She'd always imagined her life would end with a bang rather than a whimper. She'd never imagined dying so helpless, so alone. What a way to go, trapped in a tunnel under a radioactive ruin without any idea why she was there, what went on in the complex, knowing only that everyone had betrayed her. Gregor, Laurent, Frank, Karel and now – and in a strange way this hurt the most – Camilla.

She closed her eyes. It could be worse; at least she wasn't letting anyone down. There was no one waiting for her out there, no one depending on her. Karel was just a man for hire. Gregor? She thought she loved him once. She even married him. But she couldn't give him what he wanted.

When had her husband started sleeping with other women? Was it to punish her for her refusal to contemplate motherhood? Was it calculated carelessness to leave evidence of his affairs? But his *secretary*? How predictable. How pathetic. Pick someone you have power over. He lacked even the imagination to find a worthy rival. How lazy not to even leave work to fuck. Was Gregor any better than slimeball Frank? She hoped the Russians were teaching him a lesson up there.

Men weren't all bad. It was her taste in men that was the problem. Beautiful, fit, sensitive, introverted young men made her go weak at the knees. Alpha male arrogance and bombast was an instant turn-off, maybe because she preferred to hunt than to be hunted, to retain control.

It wasn't just about sex. She needed intimacy and human contact, warmth and scent, new cadence and fresh tastes. She wanted to hold and be held, give pleasure as well as take it, talk deep into the night and deeper into the soul. Jaq was a loner, but she was rarely lonely.

So why did she pursue relationships that couldn't last? Did she unconsciously seek out built-in obsolescence? Was she attracted by the fatal flaws? Did she need a get-out-of-jail-free card with every commitment, a Bickford Fuse to ensure the relationship remained time-limited, never constraining?

Was that why she refused Johan all those years ago? Because he was perfect? The problem was hers; she never felt worthy of his love, and bad things always happened to the people she cared about. Johan was too good for her, generous and uncomplicated. He deserved someone good, someone like Emma.

Would her life have been different if her baby had lived? She never told anyone who the father was. Not her mother, not the school, not the priest, not the nuns. Even when they threatened her. She'd have told Aunt Lettie, but she was the one person who never asked. What if the baby had looked like Mr Peres? The same twinkling eyes as her chemistry teacher. Would they have guessed then? The nuns didn't let her see her son. Was he deformed? The nuns prayed for his soul and assured her it was for the best. Did the sins of the mother twist and warp the baby? What of the sins of the grandmother? Angie disowned her own child and then . . . no, best not get angry.

Aunt Lettie whisked Jaq away, brought her to England, found a school that taught girls about contraception and self-confidence, allowed her to start over again in a place where no one judged her, where no one knew about her past, her mistakes. Where she became only the sum of her present actions. From the moment she left the convent, each new day was a gift.

What had Elena said about Sergei? That he wasn't brave, he just had no fear. Perhaps she and Sergei weren't so different. She wasn't naturally brilliant, but she'd worked night and day to fill the emptiness. She wasn't particularly sporty, but she always kept active. She wasn't particularly adventurous; she just viewed risk differently from most people because she had less to lose.

If there was a heaven, Aunt Lettie would be waiting for her there with a gin and tonic, tickets to a film and a booking at a good restaurant.

Euphoria washed over her. The memories she had locked down, locked in, were clamouring for release. It was time to face up to the past and set them free. If there was a heaven, she could finally meet her own son. She was ready at last.

But first she needed to take a little rest.

Jaq lay down and closed her eyes.

Saturday 4 June, Chornobyl Exclusion Zone, Ukraine

The rattling and whooshing meant something, but Jaq couldn't remember what. An alarm? Was she late for work? She squeezed her eyes shut at the sudden flashes of light all around her. Where was she? A disco? Back at the karaoke bar in Kranjskabel? Was she drunk? That wasn't like her; she could handle alcohol. Something fluttered past, caressing her cheek. The smell of grass replaced the foetid air around her. She sat up and inhaled great gulps of cool air as a wind whistled into the stagnant tunnel. A tunnel? Why was she lying on the floor of a tunnel? In the dark? She opened her eyes. No, it wasn't dark any more. A ribbon of fluorescent tubes was winking and blinking and flashing into life. Her pulse was slowing, she was breathing more evenly now. As the oxygen levels returned to normal, her brain began to function. Danger. Move!

She crawled towards the fresh air, her limbs heavy and slow. Was the ventilation fan on a timer? Was it connected to the opening of a door? She remembered now. A trapdoor. Danger. Stop!

She stood up at the intersection of the six tunnels. The Russians had murdered Petr. Oh, Petr. She was a witness. If they found her here, she'd be shot too. Danger. Hide!

How long did the trapdoor stay open? Several minutes. If she hid in one of the side tunnels, then whoever was coming in would walk straight past her, and she could double back and escape the way they came in.

Which trapdoor would they leave by? She had a one in five chance of getting it wrong. Pretty good odds compared to some recent scenarios. She looked around. The lights were bright now. She would be seen. There was really only one choice, the tunnel with the bend; she could hide just around the corner until they passed by. Which one was it? Think.

Her pulse quickened at the sound of footsteps. More than one person. The bright light blinded her, and she splayed her fingers, scrabbling at the tunnel wall, searching for the scratch marks.

The sound of women's voices, and laughter. Were they coming into the complex, or leaving? They were getting closer.

Her hands found the marks – four scratches. No! Which way, clockwise or anticlockwise? She went left, five scratches. *Merda!* The other way. Two scratches, this one.

She raced down the tunnel and reached the bend as the shadows crossed the intersection. A radio crackled.

'*Vidkreetye looyuk shist.*'

Shist. That was a number. *Odine*, *dva*, *tree*, *cheteerye*, *piyat*, *shist.* Six, *shist* meant six. Trapdoor six. So that is how they got the doors open from inside, by radio.

'*Parole?*'

She could hear the women repeating the word *parole.* Parole, as in release from prison? *Parole* as in word. Password. There was a short conversation. Everyone talking at once.

'Zagrovyl.'

Of course. She allowed herself a grim smile. She could hear a motor whirring. The familiar creaking and squealing of neglected machines. Did these shady operatives not appreciate the importance of regular oiling and greasing? They weren't too hot on tunnel ventilation either. Perhaps they needed a decent engineer. She peered round the bend; there was no one in the intersection. She moved stealthily towards the crossing. Which way had they come? The breeze carried a scent of apple. She lurched towards it, away from the retreating footsteps.

As she neared the ladder, she realised something was wrong. There was no light above it. Her heart beat faster and she began to run. Had it closed already? No, surely not, she'd have heard it. What time was it? It must be dark outside. A gasp of relief. The trapdoor was definitely open. The wind was whistling through it. Oh God, whistling. Whistling because the space was getting smaller, the same volume of air passing through a narrowing cross-sectional area at a higher velocity. She could feel the cool air, smell the grass, see the stars through the opening. Quick. The opening was closing. The trapdoor descending. At the base of the ladder she threw her backpack ahead and climbed, pushing the Tardis bag up to jam it into the closing hatch. Clang! Too late.

Jaq thumped the ladder with her fist, furious. Just a few seconds earlier and she'd have made it through. She refused to die in this tunnel. Her only chance was to run back to trapdoor six and follow the women going in the opposite direction. Maybe she could slip out behind them; maybe she would have to surrender to them. One thing was certain: she would not survive a night in this tunnel.

She yanked her bag free, tearing the backpack straps where it had been pinched between the trapdoor and frame, and sprinted back down the tunnel. In the harsh fluorescent light, the last pair of shoes disappeared up the ladder.

No time to be clever, no time to hesitate and miss the opportunity. She threw herself at the ladder and climbed, ignoring the cuts in her hand, the ache in her shoulders. Her head emerged, but she didn't have time to take in her surroundings. The motor was whirring, the hatch pivoting. With a superhuman effort she hauled herself up the last rung and threw herself through the gap. Not a moment too soon. The trapdoor sprang shut with a resounding clang.

She lay face down in the dirt and waited. Expecting the muzzle of a gun in her back. The shout as someone raised the alarm. She took slow, deep breaths, glad of the fresh air. She counted to sixty and when nothing happened, she rolled over.

And gasped.

Saturday 4 June, Chornobyl Exclusion Zone, Ukraine

Jaq looked up in amazement. The outside of the complex was 1970s Soviet brutalism gone to seed – maximum space for minimum concrete, lowest common denominator. Inside was a modern fairy tale of flawless engineering.

The trapdoor opened between two large tanks. The group of women who had climbed out of the tunnel ahead of her were already on the other side of one – they would not see her even if they glanced back. The sky was dark and there was no moon, just the tiny pinpricks of stars overhead. Floodlights illuminated several large metal cylinders, but the trapdoor lay in shadow. She rolled over towards the tank bund, a low concrete wall that divided one tank from another, counting sixteen stainless steel tanks, plus a further four wrapped in corrugated aluminium cladding. Each tank stood about six metres tall and two and a half metres wide. She did a quick sum in her head. Thirty tonnes of storage each if filled with water. Judging by the foam nozzles and toxic sensors, they did not contain water.

She squinted back towards the trapdoor. Hard to see where it was now. A helicopter stood between the tanks and the concrete buildings facing the perimeter. Now, inside the complex, she could see the buildings were follies, left to appear deserted and crumbling to the outside, forming a protective wall around this high-tech facility.

Metal tubes ran along a pipe bridge from the tank farm towards the wall and then into the ground. There must be a pipe tunnel as well as a people tunnel. Of course! The abandoned petrol station on the road from Belarus. The incongruity of the shiny new pipes and snap-shut couplings explained. It must be a pumping station.

Tankers drew up and discharged to the complex two kilometres away. The rest came in by helicopter.

She edged round one of the tanks to gain a better view of the heart of the complex. Wow. This was what she'd glimpsed when she first rolled over. Lit from inside, it sparkled, a riot of colour. Her eyes were drawn to a huge glass assembly spanning the whole five-storey building. A refluxing waterfall of golden liquids bubbled and frothed. From the neck of a giant round-bottomed flask, a tall vertical glass column led into a fat horizontal condenser wound with helical coils, a bright blue coolant filling the pig's tail. All the equipment was made of glass, but on a scale she had never encountered before. She half expected to see a giant stroll into his laboratory. It was bang up to date, and no expense had been spared. But what was it for?

She started with the tanks. They were labelled with international hazard symbols: a skull and crossbones, a dead fish, a flame. Not drinking water, then. Somewhere there must be a chemical name. She kept to the shadows and moved round each one, sniffing the air, trying to identify the faint smells. Nose-clearing, astringent. Ammonia? Sweet, sickly, organic. Chloroform? Could be useful. That fishy smell. Definitely an amine of some kind.

Three possible uses. Pharmaceuticals. Pesticides. Or chemical weapons. Given the security and the location, her money was on the latter.

Voices. She flattened herself against a tank, keeping to the shadows which lengthened with the flare of a match as a man lit a cigarette. He threw the match onto the ground where it continued to burn. She edged away from the tank. Someone needed to have a chat to this organisation about process safety.

As the man strode across the courtyard talking into a phone, illuminated by a security lamp, she recognised him. Tall, red hair, short beard. Redbeard – one of the Russians who had tried to abduct her in Kranjskabel.

'Frank won't cooperate,' Redbeard said in answer to some

question. 'Right now, he's considering his options in the basement, cooling off.' He laughed. 'We stuck him in the freezer.'

Frank would be glad of a pearl-buttoned waistcoat tonight.

Redbeard came towards her. She held her breath. Surely he wouldn't venture too close to a tank of flammable solvent with a lighted cigarette?

'Yes, boss, he'll talk all right.' Redbeard stopped and puffed. 'We'll dispose of what's left.' He laughed and spat.

Frank was going to be tortured and killed. Couldn't happen to a nicer guy. Unless she broke him out. Would he do the same for her? No chance. So what? She wasn't planning on adopting Frank Good's ethical code any time soon.

Redbeard cut the call and returned to the glass-fronted building. Jaq pressed back into the shadows.

'*Odna!*' he shouted into a panel next to a steel door.

The intercom crackled into life.

'*Parole?*'

'Zagrovyl.'

The door opened.

That same password again. *Odna*? Another number. *Odna* was the feminine version of one. So not the man. Door, *dvyer*, was it feminine? Must be.

She was debating her next move when a shaft of light streamed out from an old building to the left of the production palace. A man, bareheaded and dressed only in a loose blue boiler suit emerged backwards, pulling a large plastic box on wheels. Just a normal worker. Mario's henchmen were dressed in jeans and T-shirts, not workwear. Someone needed to have a chat to the thugs about occupational hygiene as well as process safety. Wisps of steam curled around Box Man, and she caught the scent of soap. She waited until he crossed the courtyard before creeping to the door he'd left open. Jaq peered inside; rails and rails of boiler suits hanging eerily lifeless. Behind them, industrial washing machines hummed; coloured garments and yellow foam tumbled behind circular glass doors. She slipped through the door.

Inside the laundry she kept to the walls until she'd checked it was empty. A serious operation, too big for one man, it would take a team of workers to clean and press all the dirty workwear piled up in great heaps. Break time? Night shift – minimum manning? One thing was certain, Box Man would be coming back; he'd jammed a wooden wedge under the door to keep it open. She peered round the door; the courtyard was empty. She kicked out the wedge, let the door close and lock.

Jaq turned and examined the laundry. The colours of the workwear gave some clue to the organisation. Most of the garments were a dark grey colour – presumably for the operators. She'd seen the laundryman in a blue boiler suit; a few were on the next row, for cleaners and labourers. Half a row of green ones – in the unwashed pile these were the dirtiest, the imprint of a greasy chain, two small burn marks – for the fitters and sparkies. And finally, on individual hangers, a better-quality fabric in red – obviously for the managers. She hesitated; she needed to be as inconspicuous as possible. Obvious choice was the grey, but would there be female operators in this factory? She rifled through the rail. All shapes and sizes in everything except the red. This was Eastern Europe; women did just as much heavy manual work as men, but by the looks of it rarely made the top jobs. *Plus ça change.* She selected a grey boiler suit and held it against her body for size. Should she put it on over her jeans? She checked her dosimeter. High. Her filthy clothes were contaminated, it was a warm evening and she needed to be able to move quickly. She removed her passport and stripped to her underwear, pulling on the scratchy polycotton workwear. She stuffed her clothes, water bottle and portable Geiger counter into her Tardis bag and hid it under the pile of washing furthest from the machines.

A noise at the door told her Box Man had returned. He rattled the door and cursed. His footsteps retreated. How long before he found a key? She couldn't take any chances. Was there another way out? No sign of another door, but at the far end of the laundry there was an L-shaped leg ending in a counter with a small shuttered window. A hatch connecting the laundry to another building? She slid the bolts and pulled the steel shutter up a few inches. The room beyond was in darkness, but the synthetic apple smell told her there were showers.

Jaq had experience in the design of industrial laundries. Bitter experience. The introduction of a modern laundry and amenity block at Seal Sands, the enforcement of industrial hygiene rules, was a contributing factor to the deaths.

If this was like Seal Sands, or any other factory she had worked in, then the laundry would link to changing rooms, which in turn would connect directly to the production building. The workers, like the ones she'd followed in the tunnel, would arrive in their own clothes, put them in lockers and dress in clean workwear from the laundry before going to work. Which way to go – back the way she came or deeper into the complex?

Click. The noise of a key in the door. Box Man was back.

The hatch was her only way out. She sat on the counter and swung her legs through the opening, lifting the shutter just enough to squeeze through. On the other side, she closed the shutter as quietly as she could and remained crouched on the floor, breath coming fast, heart thumping, waiting to see if the laundryman would raise the alarm.

She counted to one hundred, listening for sounds of pursuit. Box Man must have assumed the wedge had slipped, hadn't noticed the hatch was unbolted. Not yet, anyway – time to move.

Two doors, one at either end of a wide corridor divided by slatted wooden benches. Along one wall there were grey steel lockers. On the other side were white tiled cubicles: toilets and showers.

As she approached the door, voices grew louder from the other end. She slipped into one of the cubicles. Just in time. A noisy group entered the changing room, women by the sound of them. She peered through the keyhole, but her line of sight was obscured by the coats hanging on the upstands between the benches. The hatch opened with a crash, and an angry male voice bellowed something though the gap. Box Man from the laundry shouting at the women about the unlocked hatch.

If this group was leaving, then perhaps the people she'd followed were the incoming shift. How many? It had been difficult to tell in the tunnels; she shivered at the memory. Ten, twelve? Could she follow the outgoing shift, blend in with them? No, she was wearing the wrong clothes. Stupid!

The hatch closed with a thud and squeal of bolts. No escape back that way. The hatch was locked from inside the laundry again. What to do?

The door of the cubicle flew open. A woman stood in front of her in her underwear, hands on hips. Her voice was loud, angry. Jaq pressed herself against the back wall. Another woman appeared beside the shouting one and started laughing. It was even more terrifying. And yet familiar.

'*Dobry vechir miy malen'kyy inzhener-khimik!*' Good evening, my little chemical engineer.

Of course! The stance, the beehive – so this was where the superintendent from the dormitory in Sector Twelve really worked. Not in the legal New Safe Confinement project. But in the illegal chemical weapons complex. No wonder she'd been pleased to hear of Jaq's profession. She thought Jaq was destined for this clandestine factory, too.

The first woman threw up her hands and left, cursing loudly. The superintendent reached forward and grabbed Jaq by the grey lapels of her boiler suit and frogmarched her to the opposite end of the corridor, away from the exit, before using her elbow to press the intercom.

She shouted into the white box. '*Cheterya!*'

'*Parole?*' came the reply.

The superintendent nudged Jaq in the ribs.

'Zagrovyl,' Jaq said.

The door opened, and the woman pushed her forward.

Jaq blinked and shielded her eyes. It was so bright inside. The golden liquid bubbling in the tall glass column cast an unearthly light over the rest of the production hall.

The superintendent handed Jaq a hard hat and safety glasses; she put them on without protest.

The door slammed shut behind them.

Jaq was in the heart of a fully functioning chemical weapons complex, expected to work.

Perfect.

Slap bang in her comfort zone.

Saturday 4 June, Chornobyl Exclusion Zone, Ukraine

Jaq hung her head and remained silent while the superintendent delivered a stern lecture in Russian. After a while her tone grew kinder. Then the instructions started. Jaq couldn't follow everything; she focused on the pauses, the spaces between the words. When a silence followed an upward inflection, Jaq made eye contact and nodded. After a while the woman smiled, pinched Jaq on the cheek and gestured for her to follow.

The warehouse lay between the changing rooms and the production hall. Blue and yellow adjustable racking, like the Meccano she played with as a child. Pallets of bags and drums stored six high. Hundreds of tonnes of high-hazard material, a few bearing the Zagrovyl label. She stopped at one. The top bags had been removed and the next bags down slit open. Inside each bag were smaller packages, tight bricks of powder of different colours, each one labelled by hand. That explained the lumpy bags at Snow Science all those weeks ago. Lumpy because they were filled with packages of neurotoxic chemicals. It also explained the difficulty of sampling. And the confusing analysis. She'd been right all along.

And then it hit her, the genius of locating a factory here, right inside the contaminated zone. Controlled chemicals were being diverted to this complex. The kind manufactured with tracing agents to track them to their destination. If the tracers were radioisotopes, then sending them into the most radioactive location on earth would disrupt the normal tracking.

A sharp elbow in the ribs. The superintendent shouted at her to move. They continued past a row of generators, through a control room where a row of men in red overalls sat at large screens. A Roseboro distributed control system, the latest model, no expense spared.

The superintendent opened the door for her and Jaq stepped into the reactor hall. *Meu Deus*, even more impressive close up. Whoever had designed this facility did not lack ambition. And was partial to glass. A chemist then, not a chemical engineer.

They passed through an airlock into a changing room. The superintendent selected a chemical protection suit from a rack and ordered Jaq to step into it as she delivered further instructions. The self-contained breathing apparatus inflated the suit from inside. Once Jaq had snapped the cuffs over rubber boots and gloves, the hissing stopped. Inflated, protected and ready to roll.

On the other side of the airlock, a dozen workers – all in similar protective suits – worked at benches. The glassware was smaller here, where the concentrated final products were tested and packed.

As Jaq watched, one of the workers filled a bottle with liquid from a dewar flask in a fume cupboard. Thick white smoke curled upwards. Interesting.

The superintendent was watching her through the window of the airlock. Jaq turned away and tried to follow the instructions. Bench Five. Quality control. Shadow the worker. Watch and learn. Right next to the fume cupboard of interest.

Jaq waited until Blonde Beehive disappeared from view. She moved slowly towards the fume cupboard. Bingo! Her guess was right: the chemical symbol on the label was $TiCl_4$. Titanium tetrachloride. Highly hazardous.

Well, at least she had a weapon now.

Saturday 4 June, Chornobyl Exclusion Zone, Ukraine

Jaq marched towards the fume cupboard and grabbed the dewar flask. Ten litres. Just the ticket. More than enough for what she had in mind.

As a visiting industrial lecturer, she'd once taught Chemical Engineering 101 to first year undergraduates. The advantage of the profession is a transferrable set of skills, a universal toolkit. One chemical plant is much like another. Raw materials are transformed into more valuable products. Sometimes the steps are simple, like making a cake – weighing, mixing, heating, cooling, separating and packing. Often there are more sophisticated unit operations, like in brewing – fermentation, filtration, drying and distillation, but reaction is at the heart of any transformation; elements separate and recombine to form new compounds.

In less than a second, Jaq had decided what to do. Create a diversion. Easy. A little longer to figure out how to do it without killing everyone outside the airlock. She must spare the workers. Harmless dupes. People like the superintendent, just a gay girl with a beehive, in need of a job. They came in and out of the zone, squatted in the free accommodation in Sector Twelve, pretended to work on the New Safe Confinement project and were paid extra to keep quiet about what they really did.

Outside the airlock she had seen safety showers, eyewash sinks with running water, neutralising spray bottles of both citric acid and ammonia solution, toxic refuges and a rack of self-contained breathing apparatus. The alarm buttons were colour-coded according to international standards, with labels to confirm. A big red mushroom – *Огонь*, *Ogon* – for the fire alarm, a black and yellow button – *токсичный*, *toksichnyy* – for toxic. Safety first, even in an illegal chemical weapons complex.

The workers wouldn't be the problem. The ones at risk were Victor and Anton. A man who threw a lighted match into a flammable tank bund was a liability. Security men, thugs more comfortable with guns than with chemicals. The very people she wanted to eliminate.

Jaq was no stranger to chemical warfare. After all, she had punished Mr Peres for his faithlessness, deprived him of his livelihood and access to vulnerable children by swapping potassium for sodium in the school chemistry store. But since then, Jaq had spent her professional life teaching people how to avoid accidents. Now she was about to cause one. All over again.

Saturday 4 June, Chornobyl Exclusion Zone, Ukraine

Jaq made her way towards the airlock. Walking slowly, purposefully, looking as if she knew what she was doing. The workers in their spacesuits made no attempt to stop her, engrossed in their own activities.

Jaq passed through the airlock back into the reactor hall and hit the red button for the toxic alarm. The siren blast was muted – the plant had been designed not to draw attention to itself outside the walls – but the workers reacted immediately. As the emergency alarm flashed overhead, they scattered, rushing towards the toxic refuges.

She opened the flask and poured some liquid into each of the eyewash sinks. A cloud of white followed her through the reactor hall. When she looked back, she could no longer see her hand in front of her face.

Titanium tetrachloride, $TiCl_4$, forms a dense white cloud on contact with water. So dense, a few drops cause a total white-out. And Jaq had used more than a few drops. The whiteness comes from titanium dioxide, but the reaction of chlorine and water gives hydrochloric acid. Deadly to living tissue.

Jaq grabbed an escape set and a bottle of ammonia solution and headed for the basement.

It wasn't compassion leading her towards Frank, it was simple logic. Two stood more chance than one against Victor and Anton, and she needed his testimony.

At the bottom of the stairs, an archway led to a corridor with a row of storage cells. A bunch of keys dangled from a meat hook. She seized them and unlocked the first door.

The animal stench of blood and excrement made her recoil.

Merda! What went on in here? Hooks in the ceiling, pulleys and chains, knives, a whetstone – more like an abattoir than a prison. No sign of Frank, just a heap of bloodied rags in one corner.

She opened one door after another, cursing until she found the giant freezer.

Frank was on his feet the instant she opened the door. His hair and eyebrows were white, but he was alive. He stared at her without recognition. Perhaps as well.

Whoomph, whoomph. A helicopter rotor starting up. Victor and Anton bailing out. Quick. The only way out was up.

She threw him an escape set and gestured for him to follow. Back down the corridor, up the stairs. The fog swirled, dense and white. Jaq cleared a path, spraying ammonia solution, the base neutralising the acid, the mist sinking and then rising behind them as they moved forward.

A dark shape jumped off the chopper footplate and barrelled towards them. As he came closer Jaq recognised him – Bouncer, the other Russian thug from the Kranjskabel kidnap. Redbeard must be at the controls. Frank was fast. He lunged forward and rugby-tackled Bouncer at the knees, bringing him crashing down onto the concrete pad. Thud, thwack, crunch – Frank pounded his torturer with punches and kicks. Jaq yanked at the handle of the helicopter door, spraying ammonia solution into the cockpit. Redbeard screamed, twisting in the pilot's seat, his fists bunched up, rubbing at his eyes.

She pulled off her hood and vaulted into the small space. 'Frank,' she shouted. 'Get in here!'

Frank appeared at the door. 'Jaqueline,' he said. 'What a pleasant surprise. Can't keep away, eh?'

He punched Redbeard in the gut as he leapt onto the flight deck. Something metallic clattered to the floor.

'Can you fly a helicopter?' Jaq asked.

'No,' he said, and raised his fists. 'A light plane, yes.'

'Then stop hitting the pilot.'

Frank reached down and retrieved Redbeard's gun. He pressed

the barrel against the back of Redbeard's neck. '*Vamos*,' he snarled. 'Chop-chop. Fly.'

Redbeard's eyes darted around, beads of sweat forming on his brow, more afraid of the swirling chemical fog than the gun. He began to cough, flicking switches faster now, glancing over at her as he manoeuvred the controls. The helicopter strained and slowly lifted. Within a few metres the air was clear again, the moon shining from a sky scattered with stars. Jaq released her mask and breathed in the night air. They hovered above the complex, the white mist roiling, tendrils creeping over the walls.

The workers would be safe in the refuge, but if Frank hadn't killed Bouncer, then the $TiCl_4$ vapour would. He would breathe the gas; it would react with the moisture in his throat. His body would fight back against the hydrochloric acid, the tissues releasing water until he drowned in his own fluids. She shivered. A horrible way to die.

'Go!' Frank struck Redbeard with the barrel of the pistol. 'Kiev.'

'Careful,' Jaq hissed. 'We need him.'

The helicopter rose and banked. The New Safe Confinement cranes blinked below, the cooling towers of the abandoned reactors casting moon shadows over the huge artificial lake. Somewhere in those dark inkblots, giant carp swam under the still water. Fish with no predators and ample food.

The forest stretched out below, a dark tangle of trees full of deer and boar and lynx and bear and tiny horses. Stretching for miles and miles. Lights here and there. Signs of human habitation, illegal resettlers. Why would someone choose to live in a contaminated zone? How bad did life have to be elsewhere to force people to settle here? Did they understand the risks? Did anyone really understand risk? Humans were the masters of self-deception. Was it any different from choosing to smoke, to drink, to ride a motorbike, to ski? Where the risk was notional, the effects delayed, didn't humans always choose the easy way? Why else would anyone even consider nuclear power? With waste that would remain lethal for tens of thousands of years and no idea how to treat it?

Jaq sighed as the Chornobyl complex faded from view. The stars were brighter now. The Little Dipper, the Big Dipper, the Pole Star. Wait. Which way were they heading? Surely Kiev was to the south. 'Why are you going north?' she said.

'I said Kiev.' Frank pushed the barrel of the gun into the pilot's neck. 'You double-crossing bastard,' he snarled. 'Kiev, not fucking Minsk.'

Redbeard tightened an oxygen mask around his mouth and flicked a few switches. Nothing happened. Why weren't they changing direction? Did he think Frank wouldn't use the gun? Jaq had no illusions. If Frank shot the pilot, could she land the chopper? She'd travelled in one often enough going out to the rigs in the North Sea. Often enough to know that it took real skill.

The blades whirred overhead. Whoa! Her hand flew to her mouth at a sudden spell of nausea. What was the noise? Was she imagining the popping, fizzing noise? Inside her head or out? Down, look down. Coming up from the floor. Something was hissing: a cylinder under Frank's chair. Oxygen? That wouldn't cause her to feel so slow and woozy. She tried to open her mouth to call out to Frank, but a sharp metallic taste caused her to retch. She attempted to jerk to the side. With a huge effort she turned to Frank. His head was thrown back, his face white, his mouth open, gasping like a goldfish. He was brandishing the gun, waving it wildly, his eyes rolling in his head.

She managed to cry out. 'Frank, no!'

Too late. The bullet tore through the cabin and her stomach lurched as the helicopter spun. Hurtling, tumbling out of control, plummeting towards the ground.

That was when she remembered she could fly. She laughed out loud, kicked open the door, spread her wings and jumped.

The last thing she remembered was the wind in her hair and the starlight in her eyes.

So beautiful.

Saturday 4 June, Chornobyl Exclusion Zone, Ukraine

The mound of bloodied rags began to move. Very slowly the heap rose and uncoiled into human shape.

The punishment cell in the complex had no windows, but the door was wide open and the corridor lights flickered as a white mist advanced. The bloodied rag monster crawled towards the light. And then recoiled.

Boris tasted acid on his tongue before his eyes focused on the tendrils of white snaking into the cell. A chemist, it took him only a moment to place it: $TiCl_4$.

Despite his injuries, he moved fast. Yanked the key from the door, kicked it shut and retreated to the stinking corner where he'd lain for days. He placed a urine-soaked rag over his nose and mouth and coiled back into a ball.

He'd recognised Silver the moment she threw open his cell door. Was it Silver who'd released the Tickle? Silver who'd opened Pandora's Box? That was the trouble with chemical weapons, the consequences were so much worse than you could ever imagine. Hard to believe that anyone so smart could be so stupid.

Oh yes, Silver, well done, you just unleashed your worst nightmare.

PART V: GAVOTTE
POLAND TO BELARUS

Tuesday 5 July, Terespol, Poland

Boom-boom-a-boom, thump. Jaq kept her eyes closed to avoid the dizziness, and experimented with her limbs. Her muscles ached as she moved her arms against something soft and smooth. Cotton bed sheets? She tried her legs. No restraints. So why had they stuffed her mouth with cotton wool? She brought up a hand to her face and curled a finger to hook it out, but there was nothing in there.

'Water! Please.'

Her mouth was so dry, words came out as a whispered croak. Footsteps approached: rubber soles on a polished floor, not boots on concrete. Where was she? A hand supported her, cool and firm. Guiding her head forward until her parched lips touched the rim of a glass. Water. Tepid and chlorinated, it had never tasted more delicious. Someone was holding her up with one hand, controlling the angle of the glass with the other and talking, soft and low. She didn't understand the words; all her attention was focused on getting more water, fighting the rhythm of unsatisfactory sips, trying to gulp down the precious liquid as fast as possible. More! Why were they taking the glass away? She screamed in frustration as the room spun and blackness closed in on her.

The noise woke her. *Thump-thump-a-thump.* A strange sensation from her feet, her heels drumming against the mattress. She opened her eyes and fought the giddiness, raising her head to squint at the end of the bed. It was hard to focus, the room was blurred, but there was nobody near; no one was forcing her legs to make those rapid, jerky movements, and yet she was powerless to stop. The mattress was soft and the little *whoomph-whoomph* noise her pounding feet made barely travelled. But her laughter did. It sounded peculiar even to her own ears. A cackling, dry laugh, high and raucous. Her face was wet. Why? She squeezed her eyes to find that the tears were her own.

Her stomach cramped and spasmed, and she retched over the side of the bed.

A blur of white resolved into a person. Feet first, then a mop swished across the floor. White trousers, white tunic with blue braiding, clean-shaven face. He was young, his uniform more military than hospital. When she lay back, he cleaned her face. The damp towel was cool and welcome against her flushed skin. He was talking to her, low, quiet words, but they made no sense. She tried to remember her Ukrainian.

Voda – water. It came out as *vodka.* The harder she tried to correct it the louder the words shot from her mouth. VODKA! VODKA! She was laughing out loud again, her heart fluttering, completely elated. She was alive. Thank you. *Spasiba.* SPASTIC! SPACESUIT! SPAM! She roared with laughter, her whole body shaking.

He sat next to her on the bed and tapped a syringe. The tears flowed as swiftly as the laughter had stopped. She grabbed the hand preparing to administer an injection and brought it to her cheek, clinging onto it, desperate for some human contact. He was talking to her again. A lovely cadence, a gentle rhythm, the voice of reassurance. Like a cassette played backwards. Entirely free of vowels. She released his hand and waited for the prick of a needle and release.

The next time Jaq woke she was able to take in her surroundings. Hers was the only bed; a chair with a cabinet beside it and a sink opposite. She lay in a white room with high ceilings, a single door and two big windows. The bright green leaves brushing against the glass brought tears to her eyes. She bit her lip, determined not to cry.

'Good morning!'

Jaq hadn't noticed the other door, flush with the wall. A woman in a military uniform, a stethoscope round her neck, closed it behind her and advanced towards Jaq with a broad smile.

'How are you feeling today?' she asked. English. No trace of an accent.

'Where am I?' Jaq asked.

'You're at the NATO hospital in Terespol, Poland,' the woman replied.

A hospital, not a prison. 'And who are you?'

'I am Brigadier Marion Fairman. Also a medical doctor. Your doctor.'

A doctor. She was sick. That was right, she felt wretched. Everything ached.

'Can you tell me your name?'

What an odd question. 'Jaqueline Silver.'

'And your profession?'

'I'm a chemical engineer.'

'And your home address?'

Jaq glanced up at the ceiling. Where did she live? She had no idea. She stared into a dark abyss where her memory should have been. 'Doctor, what is wrong with me?'

'Do you remember what happened?'

Jaq started to shake her head and then wished she hadn't. It set off a battering on the inside of her skull. *Bat-a-bat-a-bat.*

'You have suffered a major trauma. It looks as if you were thrown from a moving vehicle. No sign of internal bleeding, but you sustained a severe blow to the head. You arrived here unconscious, and we kept you in a medically induced coma until it was safe to bring you round.'

'Will my memory come back?'

'In most cases, yes. But we need to keep you in for observation to see if there is any lasting damage.'

Why was she in Poland? Her head was stuffed with nonsense. 'What happened to me?'

'We were rather hoping you could tell us,' the doctor said. She picked up a clipboard hanging out of sight at the end of the bed and flipped through the charts. 'You were picked up in Belarus. Luckily for you, your Portuguese passport meant you were brought to us. I'm afraid you were rather distressed by the time you got here.'

'When? How long have I been here?'

'Several weeks.'

Weeks! The panic swept over her. There was something important she had to do. She shivered and closed her eyes against the cold dizziness.

'Get some rest,' the doctor said. 'Then we can talk again.' She smiled brightly.

'No,' Jaq said. 'You have to stop them!' Stop who? Where? She was crying. What was happening to her? She never cried. When had she been given a personality transplant? She accepted the needle gratefully, anything to escape the awful confusion of reality.

Jaq slipped into opiate oblivion.

Thursday 7 July, Terespol, Poland

The man from the embassy stared at Jaq, disbelief etched on his face. She saw herself reflected in his eyes. A sick woman in a hospital bed with partial amnesia. Was she mad? Had she invented this?

Was she even sure any more?

Yes, she was sure. Somewhere out there was a fully functioning chemical weapons factory. It didn't matter about anything else – her past, her future – she had to ensure the factory was closed down.

A man's face swam into her memory, sandy hair, brown eyes, his features blurred. 'There was someone with me.' But who? Think! Forest smell, damp wood, mushrooms. A scientist guide. 'A mycologist. He took me to Chornobyl.' The bus tour. No, not then. A motorbike, the open road. The gentle man who kept his body from touching hers. Petr. Of course. The bracket fungi samples, the heat of the day, the helicopters. Jaq sat up suddenly and her hand flew to her mouth. 'Petr. They killed him.'

Marion laid down her pen. 'I see,' she said gravely. 'Were there any other casualties?'

'Victor, the torturer. I killed him with tickle.'

Marion stood up. 'Jaq, maybe we should take a break.'

'No. Please, it's all coming back.'

The man from the embassy suppressed a smirk. 'All coming back, eh? Quite an adventure. You went mushroom-hunting with Petr in a radioactive zone and tickled Victor to death?'

'*Tickle!* Titanium tetrachloride. $TiCl_4$. In the secret weapons complex, I added it to water and made a cloud for us to escape.'

'You escaped on a cloud?'

'No, in a helicopter. With . . .' Oh, what was his name? 'I rescued the man who stole my Geiger counter. Well, not mine, Sergei's. They were going to kill him.'

'Who, Sergei?'

'NO!' So frustrating, so obvious. Why weren't they listening? 'The police blamed my handling of the explosives, but—'

'You handle explosives?'

'Yes, for skiing. It's my job.'

The man from the embassy coughed and stood up. 'Doctor, I think this patient needs to return home for treatment.'

'Jaq can't fly,' Marion said. 'The swelling in the brain will take time to go down.'

'I'll contact her next of kin, see if someone can come and drive her home.'

'No!' Not Gregor. The last person she needed was her ex-husband. Jaq tore at her hair. 'Johan. Call Johan! We have to stop them.'

'Stop who, Jaq?' The doctor's voice was level, calm.

She paused. Who indeed? Who were Bouncer and Redbeard working for?

'I rescued a man. He can confirm my story.'

Marion picked up her pen. 'Can you remember his name?'

What was his name? A contradiction in terms, that much she remembered. Honest Crook? Evil Weasel? Neil Humble? Bastard Bob? The long chin, the gaunt face, the thin lips curling into a sneer. After he shot the pilot, had he survived the helicopter crash?

It came to her suddenly. 'Frank Good,' she said. 'Zagrovyl. Find him.'

'Frank Good?' The man from the embassy laughed. 'The European director of one of the top international companies?' He waved a hand dismissively. 'I think I've heard enough.'

After he left, the doctor remained, her brows knitted together.

'Jaq, you need to rest.'

'You don't believe me either?'

'It's not a question of belief or disbelief. Your brain is recovering from a serious blow. Your memories are returning and you're remaking connections.'

'I am absolutely sure I was with Petr. We found a secret chemical weapons complex. The men killed Petr and were going to kill Frank. I rescued him, but we were gassed. Some kind of nerve agent made me think I could fly. I jumped out of the helicopter before it crashed. Then I woke up here. And now we have to find that complex and render it harmless.'

'What you have described is a textbook reaction to Agent Fifteen,' Marion said thoughtfully. 'It causes an immediately debilitating effect on the central nervous and parasympathetic systems followed by full-blown delirium and paranoia. However . . .'

A change in tone.

'If you had been given quinuclidinyl benzilate,' she continued, 'even if it was weeks ago, there would still be traces in your bloodstream.'

'So, we can do the tests to prove what I've been saying?'

'We have already done every test in the book,' the doctor said. 'The analysis shows nothing, Jaq, absolutely nothing.'

'So maybe there is a new nerve agent? An undetectable homologue. It was a highly sophisticated operation.'

The doctor stood up. 'Or perhaps there was no such operation. No such complex. There is no doubt you suffered an accident, a trauma. Judging by your injuries, you were in a vehicle crash.'

'Please, check out my story. Talk to the Finnish doctor at the New Safe Confinement. They treated me for food poisoning. My computer and luggage must still be in a hotel in Kiev. Talk to the tour company, see if you can find out what happened to Petr.' She choked back a sob. 'And find Frank Good.' Had he survived the helicopter crash? She gripped Marion's hand. 'Look, if there is even a small chance I didn't make this up, then you must find out what happened to Frank Good.'

Thursday 7 July, Frankfurt, Germany

Boris shrugged off his jacket and tossed it into the passenger seat. He grinned as he slid behind the wheel of a Porsche Cayenne. The organic curves of the cream leather-lined cockpit contrasted sharply with the boxy cab of the articulated lorry he was leaving behind. He wouldn't be sad to see the last of that beast. Moving on. Onwards and upwards. Oh, yes. He adjusted the mirrors, stretched his legs and paused to savour the moment, stroking his black beard as he investigated the console.

Judging by the CDs in the glovebox, the previous owner's taste in music tended to Latin funk; Boris was more of a metal head: Umbrtka, Salamandra. The square CD boxes clacked, plastic against plastic, until he extracted an acceptable compromise and slid a disc into the slot.

Turning up the volume as he accelerated onto the autobahn, he sang along to David Bowie's 'Golden Years' as the speedometer hit 200 km an hour.

Forget the sluggish response and awkward handling of a juggernaut: life was taking him somewhere at last. This was the vehicle he was born to drive.

And after this job, he would never drive an HGV again.

Boris took full credit for delivering the Tyche tracker key to Sergei's money-grubbing friend, Elena. It was a risk. A risk that had angered The Spider. A risk that almost cost him his life.

But the trap had worked. Boris was back in favour. This job came straight from the top. And what a satisfying job. Payback time.

While Mario had left him in that stinking cell, Silver had taken the bait, found the tracker. In the zone of alienation, just as he had predicted. Then stumbled into the Chornobyl factory and – he had to laugh at this – released her worst nightmare. Not Tickle – he was a chemist, he knew how to neutralise that – no, she had unlocked the very cell where he lay waiting for that bastard Mario to remember their deal.

Silver was in a bad way, by all accounts, barely conscious in hospital. Dangerous places, hospitals. Full of sick people. Anything could happen. His job was to ensure it did.

Silver was in a coma. What was the rush? He would make a detour, pick up new supplies, visit his mother. Picture her surprise when he turned up with these wheels. She'd know how to remove the bloody traces of the former owner. Imagine the admiring glances when he took her for a spin round the main square in Pardubice.

They could eat at Pavel's swanky new restaurant. *Maminka* wouldn't be able to lecture him about his career prospects this time.

Friday 8 July, Terespol, Poland

The sun streamed into the room through the high windows. Jaq lay on her hospital bed and tried to identify the birds by their songs in the trees. A cuckoo – that was easy. A woodpecker. But what was the trilling song, *la-la-la-la-la-laritzio*? It rose on the final three syllables, repeated over and over again.

Marion entered the room. An orderly followed, pulling a large television on a rolling stand.

'Well, your story checks out, in part at least.' The army doctor placed a hand on the sheet over Jaq's knee. 'There is no doubt you visited Chornobyl. The last person you were seen with was Petr, your guide. I have bad news. He died in a motorcycle accident in Belarus a few weeks ago. The bike went into a quarry and caught fire.' She dropped her eyes. 'I'm so sorry. All I can imagine is that you were a pillion passenger and were thrown off. Do you remember now?'

'No.' Jaq groaned and lay back.

'It's not surprising you don't remember the crash. Even in patients who regain almost full memory, the trauma is often the one thing that remains a blank. Maybe nature's way of protecting us from things too awful to face.'

Jaq sat up straight and met Marion's eyes. 'I don't remember a crash because it didn't happen. Get the police to investigate more thoroughly. Petr was shot in the head. The accident was staged after he was dead.'

'I'm sorry, Jaq, but there wasn't much evidence left. He wasn't wearing a helmet.' The doctor lowered her voice. 'It took some time for the body to be identified.'

Smash him up good.

'The police said no other vehicles were involved. But someone must have found you and brought you here. Just as well. Brains are funny things – we understand so little about trauma to the head.' Marion took her hand. 'Look, Jaq, as you regained consciousness, you had a series of extremely vivid dreams, nightmares even. You are a highly intelligent woman. It wouldn't be surprising if some of your dreams were . . . well, inventive.'

'You think I imagined the whole secret weapons complex in some sort of delirium?'

'Yes, and given your profession, it's not surprising your delirium centred on chemicals, an area where you are a recognised expert.'

Jaq bit her lip. It was pointless arguing like this; they were going round in circles. The more she remembered, the more specific she was, the more fantastic it sounded. Safety showers and eyewash stations among weapons of mass destruction. But why not? The criminal masterminds were not fools. Safety was good business. Injured employees would draw attention to the hidden factory. She could see the logic, but she doubted an outsider would. She needed to talk to someone who would listen, someone who would trust her judgement, believe her. She needed Johan.

The doctor misread her silence. 'I'm sorry, Jaq,' she said. 'I know this is hard to hear, but the brain is an extraordinary organ. You have lost whole weeks while you were unconscious, and then what seemed like hours of action in a dream may have been a few seconds as you woke.' She stood up. 'But I know something that might help you return to reality. You said you rescued someone?'

Jaq nodded.

'And that he could corroborate your story?'

'Yes.'

'Jaq, we found Frank Good.'

Friday 8 July, Terespol, Poland

The video-conferencing technology was better quality than Jaq was used to. It was as if Frank Good was present in the room.

'You know Jaqueline, I believe?' Marion asked.

He turned his head to stare at Jaq. Would he deny all knowledge of her, of what they'd been through?

'Yes,' he said.

The relief didn't last long.

'I know she's not to be trusted,' he said. 'She is an unbalanced troublemaker. She caused a scene at the Zagrovyl head office in Teesside and has been following me and trying to cause mischief ever since.' He waved an official document at the screen. 'I have a restraining order against her, taken out in March of this year.'

You bastard. You didn't mention that when I rescued you from your cell in the complex.

Marion dropped her eyes to Jaq's notes. 'You were in Kiev recently?' she asked.

'I had a business meeting there.'

'And in Chornobyl?'

'Chernobyl? I went on an organised tour,' he said. He puffed out his chest. 'I think it is important for leaders of industry to be informed about the world around them.' He pointed at Jaq. 'This woman has been stalking me. She followed me there.'

'And were you imprisoned in a secret weapons complex?'

Frank laughed. A sneering laugh. 'Did she tell you that?' He made a circling movement with his forefinger against his temple.

Jaq couldn't stop herself. 'I saved your life!'

Frank sat back and crossed his arms. He raised his eyebrows to heaven.

'I actually feel sorry for her,' he said. 'Is she seeing a doctor?'

'I'm her physician,' Marion said. 'She's under my care.'

Frank uncrossed his arms. 'Are you a psychiatrist?' he asked. 'Because this one is as batty as a fruitcake. And dangerous, too.'

Saturday 9 July, Terespol, Poland

Jaq sat on the side of the NATO hospital bed and waited. She stared down at the flowery plimsolls – the silver eyelets were empty, and the flaps hung loosely over the fabric tongue. Why had they given her lace-up shoes without laces? Would they give her the laces as a leaving gift? Or only when they judged that she was no longer mad?

The clothes they'd given her were mismatched and ill-fitting. Paper knickers but no bra. A salmon-pink round-necked long-sleeved T-shirt and stiff, shiny, pale blue trousers. Unfashionable, like the lace-up shoes without laces. Her eyes were drawn down again, fascinated by the shoes. What colour were the laces they'd taken out? White? Or perhaps red, to contrast with the pale fabric, a criss-cross of bloody stitches? Or yellow, like scar tissue on skin drained of blood. Did they really think she might hang herself with a shoelace? Jaq had been through many things in her life, but the thought of suicide had never entered her head. The clumsy, protective act made her see how they saw her: damaged, lost, a danger to herself.

No one believed her. Perhaps she no longer believed herself.

Stop thinking about the shoes. Think about something else.

When she was well enough, she could go home. Where was home?

Was home where her friends were? The farmhouse on the shores of Lake Coniston, where Johan and Emma always made her so welcome?

Or was it the little flat in Yarm, overlooking the River Tees? It was the only property she'd ever legally owned, and yet she still thought of it as Aunt Lettie's. Or was it her rented bedsit in

Slovenia? Home had certainly never been the grand house Gregor bought, even if she'd fooled herself into believing they were happy there for a time.

Gregor wouldn't have been seen dead with her wearing the salmon and pale blue lunatic costume. On the plus side, if he'd been coming to collect her he'd have provided something beautiful for her to wear. Although someone else would've been sent out to buy it. What were secretaries for?

Don't answer.

On the downside, he'd have found a psychiatrist and mental asylum to have her committed the minute they arrived in England. Perhaps that was Frank's plan as well. For Gregor it was avoidance of inconvenience disguised as concern. For Frank, it was more threatening.

Or was home with her mother in Lisbon? Should she be back in Campo de Ourique, sidling past the convent every day, blocking out the memories of her own incarceration there to visit her only living relative in the dementia block, unrecognised and unwelcome?

Was home in Angola, the country of her birth, a country torn apart by a civil war, a civil war that stole her brother, her father and the soul of a mother who fled, abandoning her children to save herself?

Going home? She had no home. No point in pretending. No point in anything any more. Nothing left. No one who cared. No hope.

Her eyes were drawn back to the shoes. Why remove the laces, when she could perfectly well hang herself with a sleeve of the T-shirt? Or a leg of the trousers? She was an engineer, for Christ's sake. If she wanted to make a noose she would find a way.

Saturday 9 July, Terespol, Poland

Emma appeared at the doorway, her fair curls backlit by the sun. The scent of rose water followed as she bustled into the hospital day room, swooping to retrieve the plastic beaker dangling from Jaq's hand as she woke from a doze.

Jaq allowed herself to be embraced by Johan's wife, turning her face away to hide the disappointment. Stop it. Smile. Any friendly face was welcome.

'Good gracious! You look terrible,' Emma exclaimed. 'Are you fit to travel?'

'The doctor says I can't fly yet.'

'Not dressed like that, you can't.' Emma wrinkled her mouth in disapproval.

Emma cared about her appearance. Naturally short and slight, fair and freckled, always neat and tidy. Today she wore a light checked jacket, a crisp white cotton shirt, beige trousers and flat black boots, polished to a shine. All she needed was a crop and hat and she could have galloped straight from the stables. By contrast Jaq resembled an asylum inmate, tall and scrawny, dressed in hospital clothes and lace-free plimsolls.

Jaq chuckled. Emma covered her mouth and turned away. A snort of amusement rising to a full-blown, belly-aching guffaw. Jaq joined in, her diaphragm heaving and eyes wet with mirth. An unfamiliar feeling, that tightening of the larynx from joy rather than fear.

Jaq composed herself. 'How are the kids?'

'You're wondering why I'm here instead of Johan.' Emma squinted at Jaq from under her lashes.

'I didn't mean to—'

'It's okay, I understand. Johan is in the middle of a major contract, a corporate boot camp. His parents came over to help. Jade and Ben are fine with their grandparents. I wanted to come and get you. We all have skin in this game, remember?'

'Emma, I'm so sorry—'

Emma put a finger to her lips as the door opened and a woman entered, accompanied by a young man, both in uniform.

The woman extended a hand. 'Brigadier Marion Fairman.'

Emma shook her hand. 'We spoke on the phone.'

Marion gestured to the man beside her. 'This is our legal adviser, Major Thomas.' The young man in uniform bowed, keeping his hands firmly at his side.

'Legal adviser?' Emma queried.

Marion's mouth pursed with apology.

Major Thomas stepped forward. 'We need to confirm Miss da Silva is not suffering from delusions as a result of her accident. Mental health is every bit as important as physical health.' He stared at Jaq. 'A condition of her release.'

'Shall we begin?' Marion unhooked the chart from the end of the bed.

'Jaq suffered severe bruising and trauma to the chest and head. The broken ribs are healing, but there is swelling in the cerebellum and Jaq is to rest and continue the anti-inflammatory treatment. She should not fly for a few months, but otherwise she is physically fit to be discharged.'

Major Thomas remained standing. 'Miss da Silva, do you remember how you sustained your injuries?'

'I jumped out of a helicopter,' Jaq said.

Emma gasped.

Major Thomas shook his head. 'Try again. You were in a motorbike accident.'

The sun shone through the window. A beautiful day. Oh, for that warmth on her face, wind in her hair. Oh to be outside, away from this place.

'I was in an accident. I don't remember anything until I woke up here.'

'Good, very good.' Major Thomas nodded. 'Do you remember who you were with?'

Frank Good. And chances were, Frank was dictating the conditions of her release. Zagrovyl must be threatening legal action if she continued with her story. The ceiling pressed down on her, the walls closing in, the air heavy with bleach, sterile, suffocating, her throat tightening. She had to escape. Tell them about the journey with Petr. Give them what they wanted.

'I was with a local guide. A mycologist. On his motorbike.' Her voice caught in her throat. 'He was killed.'

'In the motorbike accident,' Major Tom insisted.

'So you tell me.'

Emma laid her hand over Jaq's.

Major Tom continued. 'You used to work for Zagrovyl, do you remember?'

'I never worked for Zagrovyl. I worked for ICI.' Jaq tried to pull her hand away, but Emma held on firmly. 'I left when Zagrovyl bought the business I worked for.'

'Zagrovyl sacked you.' Major Tom flicked through a few pages. 'It came at a bad time. Your marriage breaking up. The inquiry over unexplained deaths at work.'

'The deaths were not unexplained.' Jaq sat forward.

'HSE vs Zagrovyl, case 4469245,' Emma said. 'Holiday death syndrome – three factory workers suffered fatal heart attacks. The men had become used to a level of nitroglycerine exposure. When Jaq modified the process to improve containment, men with undiagnosed angina died. Case dismissed. No prosecution.'

Almost any change – even improved safety – can have unintended consequences, far-reaching and difficult to predict. Jaq was vindicated, but the private prosecution, brought by the men's families, still dragged on.

'You think this is revenge?' Jaq asked. 'You think I made up a story about Zagrovyl because I was miffed about losing my job?' Jaq clenched her fists. 'I left because I couldn't stand the company, their attitude towards their workforce, their values – or lack of them – their hypocrisy.'

Major Thomas ignored her. 'While recovering in hospital you suffered hallucinations. You claimed to have discovered a secret weapons complex supplied by Zagrovyl, inside the Chornobyl zone of alienation.'

Emma squeezed her hand.

'You now recognise no such place exists. You have been cured of your delusions.' Major Thomas placed the sheet of paper in front of her. 'Here is the statement I need you to sign.'

Jaq read it and passed it to Emma.

'And if I don't?' Jaq asked.

Major Thomas frowned. 'Then I'm afraid you'll have to remain here until the treatment is complete.'

Jaq turned to Marion. The doctor met her gaze for a moment, then dropped her eyes. 'My hands are tied on this one, Jaq.'

'Emma?' Jaq pleaded.

Emma stood and addressed the legal adviser. 'This is a NATO establishment, right?'

Major Thomas nodded.

'And UN military law applies?'

'Yes.'

'Jaq, as a lawyer, I cannot advise you to sign any statement you do not believe to be true.' Emma sat back down and hugged Jaq, leaning close to whisper in her ear. 'But made under duress, it won't stand up in any court, so sign whatever it takes to get us out of here.'

Us.

Emma said *us*. A rush of guilt swept over Jaq as she admitted to herself how much she had wanted Johan here. Someone physically strong and fearless to bash the bad guys. But Emma was exactly the person she needed, deceptively soft-spoken with a razor-sharp intellect and finely honed legal mind.

Us.

Jaq had put Emma and her children in danger. She didn't have to be here, she was free to leave any time she wanted, but she had thrown her lot in with Jaq.

Emma believed her.

Jaq grabbed the pen dangling from the medical chart and signed three copies.

'Come on,' Emma said. 'Time to leave.'

Saturday 9 July, Wrocław, Poland

Boris sang along to the Bowie compilation as he drove.

The music was interrupted by a call. Mario. Angry Mario. 'Where the fuck are you?' he shouted.

Boris clenched the soft leather of the steering wheel. *Where the fuck were you while I was locked in that cell? I was right. Silver took the bait. My trap worked. You did it my way, but you left me caged. Payback isn't far away, Mario, just you wait. I'll take my time. Finish the job. And then . . .*

'On my way,' Boris said.

'You are too fucking late. She's gone.'

Not gone as in dead. Otherwise The Spider would be ringing to discuss his bonus instead of Mario chewing him out. Gone as in moved on. *Kurva*, how did that happen?

'Gone where?'

Mario exploded with rage. 'That's for you to find out, fuckwit.'

A crackle as Mario cut the call and Bowie's plaintive tones came back on the stereo.

Boris pulled over to look up the number for the hospital in Terespol. Some military stuffed shirt with a broom up his arse gave him the runaround. It took several calls before he was able to get through to a Czech orderly, who confirmed that Silver had left that morning. He asked after her health.

'Still pretty weak.'

Good.

'A friend is driving her back to England.'

Boris pulled out the atlas of Europe. Terespol to England. A twenty-hour drive to the Channel. They'd have to stop somewhere. He spun a lengthy story about driving all the way to see her, his disappointment at missing her. 'Do you have a mobile phone contact?'

'No, sorry.'

Sakra! 'I have something for her. It's important. Any idea of the route they might take?'

A door slamming, machines beeping, voices in the background as his contact consulted others.

'Hang on.' A rustling of paper. 'They asked about hotels near Łódź. One of my colleagues made a booking for them at a guest house.' He gave the details.

Silver and her lawyer were heading west. Boris was heading east. And when they met, he would finish this once and for all.

The track changed. 'Ashes to Ashes'.

Boris spun the wheel and pressed the accelerator to the floor.

Saturday 9 July, Łódź, Poland

Emma drove west. The wrong direction. Jaq closed her eyes. Plan, plan, you need a plan. To fail to plan is to plan to fail. To fail, fail, fail . . . When she woke, they were drawing into the car park of a guest house.

'We'll stop here, I'm expecting a call,' Emma said.

The guest house lay on the outskirts of Łódź. A cottage with a large garden, beautiful and productive. Mature fig, walnut and nespera trees – the large, dark green leaves still hiding some brown-spotted orange fruit – created a windbreak for a more delicate orchard. Plum, peach, cherry, laden with ripening fruit. An arch of wisteria formed a long, shady tunnel from the back door, delicate pale green fronds and hanging purple clusters of tiny flowers. The back of the house was a riot of brightly coloured bougainvillea, crimson and pink.

They paid for two rooms and asked for coffee in the guest sitting room with French windows opening out on to a busy vegetable patch.

'So, tell me everything,' Emma said.

The time for secrecy was over. She'd hit rock bottom and had no reserves left. Running on empty, it was time to put pride aside. 'I need your help,' Jaq said.

The words tumbled out – her search for Sergei, starting with his personnel files in Snow Science; the awful death of Stefan Resnik, and her incarceration in the women's prison at Ig. She skirted over the escape to Lisbon and the meeting with Elena in Minsk, describing the first sighting of Frank in Kiev, the Chornobyl tour and watermelon sickness, finding the Tyche tracker and losing it to Frank, going back into the zone with Petr. Her voice broke as she described Petr's murder and Camilla's perfidy. For the first time since the crash, the sequence of events became crystal clear in her mind. She gave a vivid portrayal of the complex, Frank's rescue, the helicopter escape and nerve gas attack.

Emma wrote everything down on a ruled yellow legal pad. A lemony scent wafted in through the open window, nature in abundance: artichokes – little crowns on long, woody stalks; beans climbing a cane trellis – bright red flowers; an unruly bunch of sorrel. Through the door of an old wooden shed, the tools – spades and hoes – stood to attention, ready to wrest bounty from nature. On a shelf lay the chemical helpers – packets of treated seed and a half-open bag of fertiliser with the familiar Zagrovyl logo.

Emma followed Jaq's eyes. 'You want the good news or the bad news?'

Jaq sat up straight. 'How bad is the bad?'

'Depends if you are Sergei Koval.'

'You found him?'

'He's dead,' Emma said. 'They finally identified the body in the Snow Science warehouse.'

Stomach twisting, nauseating recall. The bone and flesh and blood. The body blown to smithereens. She would never meet Sergei. A good guy. What a waste.

Emma extracted a newspaper report from a folder and read aloud. '*Helicopter pilot and avalanche safety expert Sergei Koval was identified from his dental records several months after a tragic accident—*'

'Accident, my arse.' Jaq smashed a fist into her palm. 'It was a booby trap.' Semtex in the coffee machine.

Emma continued. '*Sergei's employer, Snow Science, commended Sergei for the pioneering work he did on snow safety.*' She handed Jaq the newspaper cutting. A photograph of the funeral. Laurent was there, and . . . could it be Karel? No mistaking the bright curls or the way he stood, legs apart, shoulders back, chin forward. What was Karel doing at Sergei's funeral? She shook her head. There were more pressing issues to consider.

'Am I still a suspect?'

'I contacted your lawyer. He's waiting for the toxicology report on Stefan Resnik. If it shows ergotamine then we are in trouble, and you will definitely face trial in Slovenia. And the English police still want to talk to you about the drowning of Bill Sharp.'

'So, I'm about to be arrested?'

'Major Thomas didn't do his homework. If he had waited a couple of days, he could have handed you straight over to Interpol.'

'Is there any good news?'

A scuffling noise. Emma pressed a finger to her lips and flung the door open. A small grey cat bounced in and rubbed itself against Emma's boots.

She closed the door and continued. 'You remember the new contract I told you about, the corporate boot camp?'

'Yes.' Jaq reached out a finger to the cat. He sniffed and then jumped onto her lap. The soft, silky fur, the way his whole body vibrated when she stroked him, brought a temporary calm.

'Guess which company wanted an intensive programme of team building through extreme outdoor activity?'

'Not—'

'Zagrovyl. A Frank Good initiative to instil some discipline into his UK team. Your friend Frank couldn't make the first course.' Emma referred to her notes. 'Probably because he was in a freezer in Chornobyl at the time.'

Jaq smiled and tickled the cat behind the ears.

'The rest of his team didn't seem too disappointed by his absence. In fact, when Johan changed the course to better suit their abilities, a couple of them became positively loquacious.' Emma reached into her bag and extracted a slim paper binder. 'Shelly, Frank's former PA, provided full details of all the shipments to Snow Science and some very interesting information on Tyche.'

'Tyche? As in the company that made the Tyche tracker?'

'Tyche was more than that. It was a non-profit dedicated to destroying chemical weapons.'

'Why have I never heard of it before?'

'It had to be discreet. Governments had been lying for years about what stocks they actually held – mustard gas, nerve agents. Tyche was so discreet, even terrorist organisations sent material for destruction.'

'Is it still operational?'

'Tyche got into financial trouble. Zagrovyl bought them out.' Emma closed the file and held it out. 'Jaq, why would Zagrovyl want to buy a company specialising in chemical weapons decommissioning?'

Jaq leafed through the file. 'It looks as if they had special know-how, experience of handling the nastiest stuff. Zagrovyl makes medicines and weedkillers and pesticides. We wage war on disease and weeds and insects using the same basic chemicals used in chemical weapons.' She gazed out at the garden.

Emma's phone beeped. 'Here's the call I was waiting for.' She smiled. 'Look.'

Emma showed Jaq the photo as it came through on her phone. Two men on a mountain, one sitting between shards of rock, eating a sandwich. Johan smiling and waving. The other man dangling from the end of a rope over a grey abyss.

'That's the other reason Johan couldn't come. He's roped up on the top of Helvellyn. Right now, Johan is the only thing stopping Frank Good falling five hundred feet from Striding Edge. I'm sure Frank'll be happy to answer all your questions.'

Saturday 9 July, Cumbria, England

It was not one of his better decisions, he knew it now. Frank rarely made a mistake, but when he did, he was big enough to admit it. Of course, he would keep the realisation to himself; any hint of regret might destroy team morale.

The mountain was magnificent: grey, rugged and powerful. The water streamed down from the heavens, dripping over rocks into the pools, cascading from pools as little rivers. It was a long way down, first stop a hanging corrie hollowed out by millennia of ice action. The distant yapping of a dog put him in mind of a poem.

The crags repeat . . . in symphony austere . . . the sounding blast,
That, if it could, would hurry past; But that enormous barrier holds it fast.

Frank tugged on the rope. Okay, Mr Extreme Boot Camp leader, Johan whatever your name is: enough now. Pull me up.

The rest of the team had made it across Striding Edge without incident, and now their bright cagoules receded into the mist, led by the other instructor. Frank was a little surprised they hadn't stopped. Would he have waited for them? Certainly not. Just as well they had continued their ascent of Helvellyn, as there was something slightly undignified about his current predicament.

If he was to be brutally honest – and he was always brutal, although honesty was overrated and to be used sparingly – the whole corporate boot camp idea might have backfired a bit. Such a good idea, the image of fat Nicola puffing over a high ropes course, Raquel wobbling up a mountain in her ridiculous shoes, Eric adrift outside the drawing office, but they had all turned out fitter and more resilient than he expected. Who knew that Nicola was a Brownie leader? Or how much Raquel worked out in the gym? Who would have guessed geeky Eric was a keen hiker and knew the Lake District like the back of his hand?

The toll of corporate life in the fast lane – airplanes and business lounges, lavish lunch meetings and lengthy dinners – had affected his own fitness. Perhaps Bill had been letting him win at squash. He wasn't twenty any more. In his mind, this was to be his triumph, bounding ahead over the rocks while his team crumbled. In practice, they had raced on ahead and he was left here, dangling from a rope.

Frank tugged again. Johan appeared over the lip of the ridge, a long way away.

'Pull me up!' Frank shouted.

Johan shook his head and lowered something down. A small rectangular object on a string. What the fuck? He reached out and grabbed the phone as it swung down. The screen lit up.

He pressed it to his good ear. 'Who the hell is this?'

'Hello, Frank.'

There was no mistaking the voice.

Jaqueline Silver.

Saturday 9 July, Łódź, Poland

Boris bent to tickle the little grey cat brushing against his trouser leg. He'd always had a soft spot for animals. His mother had never allowed pets, but a farm cat once crawled into their attic and gave birth to five kittens. He visited every day, keeping it secret until they started moving around at night. The adults decided the scratching noises in the ceiling meant rats and were preparing to lay down poison when he told the truth. He was astonished that his mother didn't beat him. She told him to fetch the kittens from the attic. Amazed at her calm, he told her their names as he brought them down one by one – Blacky, Whitey, Spotty, Stripy, Softy – told her all about them, the ones who liked to play with his shoelaces, the ones who purred the loudest, the shyest, the sleepiest, the bravest, the softest. His mother repeated the names as she dropped them into a bucket of water. And then she beat him.

The cat's fur was soft and silky. He could feel a little heartbeat pulsating under his fingers through the thrum of purrs. When all this was over, when he'd finished the job, he'd get a cat. Maybe two. A car, a flat, a cat.

The owner of the guest house, a portly middle-aged man, appraised him from behind the counter. His smile faded as Boris put the cat down and approached, pulling on a pair of leather gloves.

'We're full.'

Liar. The place was deserted. He'd been watching it for hours. The last member of staff had just left and only one car remained in the car park. Two guests, one live-in owner and a friendly cat.

'I'm looking for my friends,' Boris said. 'Two English women.'

The man shook his head. 'I'm sorry—'

Boris leant across the counter and grabbed the man by the windpipe. With the other hand he opened the visitors' book. Nothing. He squeezed hard, lifting the man from his seat, deflecting the blows from flailing hands, before dropping him abruptly.

'You will be.' Boris vaulted over the counter and straddled the man sprawled on the ground, fighting for breath. 'Which room?'

When the owner failed to answer, Boris kicked him in the stomach for good measure. Yanking the telephone and internet sockets from the wall, he grabbed the master keys and went through the hotel, flinging open one door after another. Nothing – every room was empty.

Boris returned to the owner, still gasping like a fish out of water.

'Where are they?'

'Not here,' he croaked.

'When did they leave? Where did they go? What car are they driving?' The kicks came hard and fast, his frustration finding relief in the hard physical activity of lifting a ninety-kilo man and slamming the soft body against the hard surfaces of floor, wall, counter, fireplace, stairs.

After a while there was no point asking any more questions.

The little cat emerged from the garden. He mewed and rubbed against Boris's leg. Boris bent down and stroked it, but it didn't feel the same with gloves on.

Where the fuck were they? Somewhere between Łódź and London. *Jehla v kupce sena.* Needle in a fucking haystack.

Boris rifled through the office. He took all the cash, so it would look like a robbery. And that's when he got the lucky break. In the till, he found a booking form with a UK contact for the lawyer: Blondie's mobile phone number. It wouldn't take him long to get a trace on that.

As he was leaving, the little cat crouched beside the owner, busy with its long pink tongue.

Saturday 9 July, Poznań, Poland

Emma drove through flat, industrial landscape as Jaq recounted the conversation with Frank.

'All of a sudden, Mr Good is prepared to cooperate.'

'My husband has a way with people.'

'And how.' Jaq laughed. 'Frank is anxious to contact Major Thomas and Brigadier Fairman and correct any misunderstanding.'

'Johan won't let him off the mountain until he does.'

'Frank claims the nerve gas made him forget.'

'So, the bracing mountain air refreshed his memory?'

Emma turned in at a petrol station.

'He's going to make a full statement to the police.' As they came to a halt, Jaq kissed her friend on the cheek. 'Thank you, Emma.'

Emma turned off the engine and wrapped her arms around Jaq, squeezing her tight.

So warm and comforting. Curled up in her friend's embrace, she considered forgetting about everything. Let others resolve the mess, put things right, catch the bad guys. Emma was safety, strength and softness; she was courage, competence and compassion – the kind of mother Jaq never had. Ben and Jade were so lucky.

'I'm glad we decided not to stay the night,' Emma said. 'Let's go home.'

Go home to where? To what? To the knowledge that she had failed? Jaq sighed and shook her head. 'We need hard evidence.'

'But Frank—'

'Not good enough. He destroyed the tracker.'

'So, what do we do?'

Jaq looked across the petrol station forecourt, staring at the lilac bushes by the side of a field. She scratched her head. 'Emma, is prostitution legal in Belarus?'

Emma frowned. 'Definitely not, why?'

Jaq thumped the dashboard. 'Then I know where to go. But it's in the opposite direction.'

'Where to, partner?' Emma asked.

'Minsk.'

They doubled back, past Terespol, crossing the border at Brest.

'Emma, are there Geiger counters at border crossings?'

Emma nodded as she drove away from the border and onto an empty highway. 'The US funded a programme as part of the war on terror after nine-eleven. Not just borders – global ports, and motorway tolls and bridges, too.'

That explained the dots and ribbons on the map. Jaq imagined herself as a series of dots travelling across Europe: Teesside to Slovenia and back, Slovenia to Portugal, Portugal to Belarus, Minsk to Kiev, Kiev to Chornobyl, Chornobyl to Terespol, Terespol to Łódź and now the journey to Minsk. The dot would be a strong colour at the start, a bright yellow, opaque and full of energy, fading a little with each journey, now a transparent, weak shadow of her former self. Who cared? She was no longer alone. Her strength sat beside her. Emma represented a new, bright pink dot, moving fast.

Jaq closed her eyes. The map on the Tyche tracker followed controlled chemicals moving between Teesside and Chornobyl via Slovenia.

Why Slovenia?

'Are the rules on transport of dangerous goods different in Slovenia?'

'Let me think.' Emma tapped the steering wheel with her fingers. 'Slovenia joined the European Union in 2004, right?'

'I think so.'

'So they would have ten years to harmonise with EU rules.'

She paused to overtake a horse-drawn cart, checking her mirror, indicating left to pull out and then right to return. 'EU regulations would apply immediately for all material coming into Slovenia from the west.' She swore and braked sharply as a moped pulled out from a side road. 'But the old USSR rules would still apply for transport out of Slovenia to the east.'

'Aha!' Jaq slapped the dashboard. 'Now I understand.'

Emma hooted at a bicycle. 'Well, I don't.'

'SLYV sourced the ingredients for chemical weapons in Teesside, but they couldn't ship them direct to their secret factory because controlled chemicals contain tracers to follow their use. So SLYV hid them among shipments of explosives. The tracers set off alarms at every border and port and motorway toll station. Explosives also have tracers, so no one looked too closely. The paperwork showed a delivery to a licensed organisation, Snow Science. Tracking file closed.'

Emma nodded. 'So far, so good.'

'Then the shipment was rejected by Snow Science. New papers were prepared for the reject to go east for recycling. Different rules, no EU control of the shipment and no official notification when it vanished. And SLYV paid for everything they stole.'

Emma whistled. 'Mimicking a legitimate business, right up to the moment they purveyed weapons of mass destruction to the lunatics of the world.'

They skirted Brest and passed empty fields, wide, deserted roads, towns with no sign of shops or cafés, only the occasional playground devoid of children. In Baranovichi they stopped to buy vodka and cigarettes for Elena. The selection was limited.

'Where is everyone?' Emma whispered.

'Huge country, tiny population,' Jaq said. 'Agricultural land contaminated by the fallout from Chornobyl. Massive emigration, and those who remain live in the cities.'

As they approached Minsk, residential blocks appeared, serried escarpments of high-rise concrete. Jaq directed Emma to the

Shiskina Model Agency. The familiar red arrow pinpointing the brothel flashed and crackled overhead.

Emma raised an eyebrow and buttoned her blouse up to the neck. 'Will she cooperate?'

Jaq scowled and pressed the bell. 'We'll see.'

The steel door opened and Elena stood at the doorway, hands on hips. 'Go away.'

'Let us in,' Jaq said. 'We need to talk.'

'What do you want?' Elena glanced behind her shoulder. 'I am expecting clients.'

Jaq offered the vodka and cigarettes. Elena inspected them with a snort of distaste. 'Rubbish,' she said. 'You bring me rubbish.' She snatched them anyway and started to close the door. 'It's not safe.'

'We have news.' Jaq pushed against the door and forced her way in, Emma stumbling behind her in the dark. The gloomy anteroom was lit by a single bulb under a red lampshade and stank of cigarette smoke. Elena clapped her hands and the velvet sofas and leather chairs emptied of half-naked women. Jaq sat on a high wooden stool in front of the bar, pulling out another seat for Emma as Elena moved to collect a bottle.

Jaq splayed both hands on the smooth wooden counter of the bar. 'I'm sorry, Elena,' she said. 'Sergei is dead.'

Elena poured a splash of vodka into three glasses and slid two across the bar.

Emma wrinkled her nose and backed away, glancing sideways at Jaq, who downed her shot in one, matching Elena move for move.

'I know,' Elena said, wiping her lips with the back of her hand. She continued in Russian. 'Why tell me what I already know?'

'They came here, didn't they?' Jaq said. 'The Russians from SLYV.'

Elena blanched and filled her glass again, glancing at the door. Was she planning her escape, or waiting for someone to appear?

'If I could find you, then so could SLYV,' Jaq said. 'You were

Sergei's partner, easy to track down. When Sergei disappeared from Snow Science, SLYV came here to find you.'

'Please,' Elena said. 'Go now.'

'What did you tell them?' Jaq asked.

'I didn't tell them anything,' Elena said.

'Jaq.' Emma tugged at her sleeve. 'I think we'd better go.'

'We need to stop them,' Jaq said. 'Make sure they never come back again.'

'They are bad men,' Elena said. 'They frighten me.'

'Did they hurt you?'

Elena lowered her eyes. 'They took my girls,' she said. 'That was worse.'

'Jaq, let's go.' Emma tugged at her sleeve.

'Elena, what did they want?'

Elena hung her head. 'Boris brought a key. Said it was Sergei's, that Sergei had something hidden near Chornobyl. That the key was for a locker. Tried to make me tell them where it was. I said I didn't know. They said I had to find someone who did. Said they would follow whoever came for the key. So I gave it to you.'

Thanks a bunch. With friends like Elena, who needed enemies?

'Now go!' Elena stamped her foot and flung an arm towards the door. 'Before it's too late.'

'Not without the evidence,' Jaq said.

Elena scowled. 'What evidence?'

Noises at the door. Emma stood up. 'I've got a bad feeling, Jaq. Please, let's go now.'

'Where is it, Elena?'

Elena raised her head, her double chin jutting forward. 'I don't know what you're talking about.'

'Give me the memory stick.'

It had come to Jaq while she was describing the device to the man from the embassy. The tracker was shoebox-sized, black with a lilac sheen. It had a screen, a keyboard and a USB port.

Sergei didn't need to remove the instrument to supply the evidence; all he needed to do was download the information. But Sergei wasn't working for free, he was being paid. Follow the money. Elena was the fence, so when Elena received the money, she was meant to send the memory stick in return. But Camilla had disappeared without sending the money.

Jaq vaulted over the bar to the cash register. She yanked out the small black dongle and inspected it. It glinted with a lilac sheen.

Elena grabbed her wrist. 'What are you doing?'

Jaq smiled. 'Hide in plain sight.' *Brilliant.* She wished she'd met Sergei. Wished he'd had a happier end. She threw the memory stick to Emma. 'Let's go.'

'They promised me money for that.' Elena released Jaq's wrist and lunged towards Emma. 'Where's my money?'

Emma slapped a couple of notes on the table, enough to cover the drinks.

'You'll pay for this,' Elena screamed. 'In full.'

Saturday 9 July, Minsk, Belarus

Boris rolled up to the club on the corner of Ulitsa Shishkina. He walked round the block to check for a Polish hire car – no trace of Silver or Blondie – so he hammered on the back door. Elena flung it open and then froze, turning pale. She made a half-hearted attempt to close the door in his face, but he slammed into her, clamping a hand over her mouth before she could scream a warning.

'Hello, Elena,' he whispered. 'Nice and quiet, so no one gets hurt?'

Her eyes blazed, but she nodded slowly. He forced her back into the storeroom, turning her large, soft body to face the crates of beer while he bound her hands behind her back.

'What do you want this time?' she hissed.

'Nice welcome,' he sneered. 'Were you as hospitable to your English visitors?'

She held her breath just for a moment, long enough for him to know they had been here.

'What do you mean?' Her voice trembled.

He laughed. 'You always were a lousy actress, Elena.' She flinched as he grasped her shoulder and swung her round to face him. 'What did you tell them?'

'Who?'

He slapped her across the face. Not too hard. Just hard enough to remind her of what was coming.

'Silver. And Blondie, her little lawyer friend.'

'Why would I talk to them?'

He slapped her again, harder.

'What did you tell them?'

'I told them nothing.' Elena spat and a tooth rattled onto the concrete floor with a stream of blood and saliva. 'I know nothing.'

Boris raised his hand and then stopped.

'Any new girls here tonight?'

Her eyes widened in alarm. 'You leave my staff alone.'

Boris laughed. 'If you treat all your customers like this, I have no idea how you stay in business.'

'If you take more girls away, I will have no business,' she retorted.

'So, let's go and talk to them, maybe they remember something about the visitors.'

'No,' Elena said, squaring her shoulders. 'You talk to me.' She met his gaze with a fierce one of her own. 'I'll tell you everything.'

Saturday 9 July, Minsk, Belarus

A crescent moon hung low in a black sky. A street light flickered above the car; the shadows writhed and danced.

'Where to now?'

'Back to Terespol?' Jaq said.

Emma shook her head. 'Too late.' She showed Jaq her phone. 'The European arrest warrant was issued at noon. Birth name, maiden name, married name. They'd arrest you at the Polish border.'

'Does the EU have an extradition treaty with Belarus?'

'Nope.'

'Then let's find the best hotel in Minsk.'

The room was grand, with high ceilings and two queen-sized beds. One wall of glass, floor-to-ceiling windows three storeys above Lenin Street and directly above the imposing entrance canopy. Behind the beds were mirrors, with a dressing table between them. The other wall had a heavy wardrobe containing an ironing board and trouser press on one side of the entrance door, a chest of drawers on the other. The beige carpet stopped at the entrance to the bathroom where the thick pile carpet ran smoothly into cream tiles.

Emma plugged the memory stick into her laptop. A new icon appeared on the computer, a wheel of fortune: *1% downloaded, estimated time . . . twenty minutes.* A big file.

A knock at the door. *Merda.* Jaq's heart missed a beat.

'Room service.' A woman's voice.

'Put the chain on. Check who it is,' Jaq said.

Emma raised her eyebrows but did as instructed. She released the chain and accepted the room service trolley, barring entrance to the inquisitive waitress by means of a large tip.

Emma laid out the meal on a small table: borscht, pork meatballs in a vegetable stew, fruit salad and a bottle of wine.

'Cheers!' Emma said.

'Bottoms up!' Jaq drained and refilled her glass. Her first wine for weeks tasted so good. The faint yellow dot was growing brighter.

Outside, the government buildings were lit up against an indigo sky. The rest of the city was in near darkness. Where did their power come from these days?

Ping! *Download 100% complete.*

Emma pressed a few keys. 'I got the Tyche software from Raquel, Frank's unhappy PA,' she said. 'Damn, it doesn't seem to work.'

Blank screen, blinking cursor. Jaq knew what to do.

Login – ключ

Password – 12016834

Bingo.

A picture formed on the screen. A map of Central Europe and the Middle East. Ukraine lay in the centre, coloured dots connected by ribbons weaving through Russia to Grozny and then fanning out to Iraq and Syria, Pakistan and Afghanistan. The dots growing smaller and fainter with time. Exponential decay. Known half-lives for each radioisotope declining as time progressed, each at a different rate.

How were the deliveries made? By helicopter and then private vehicle? Regular small quantities heading for the basket case countries of the world, the flashpoints of geopolitical distress.

Jaq scrolled through the maps. Wait a minute. She clicked in increasing frustration and then threw her head back. '*Bolas!*' She slammed her fist onto the bed.

'What's wrong?' Emma asked.

'This is only half the story.'

The memory stick had no data for the first leg of the journey, the material moving from Teesside to Chornobyl via Kranjskabel.

The screen showed only the second set of maps, the routes from Chornobyl fanning out to end customers.

Proof the complex existed.

But no proof of Zagrovyl supply.

Would it be enough?

'I'm sending this to Johan.' Emma typed a few keys. 'Damn! The file is too big . . .'

A knock at the door. Heart skipping, Jaq sprang to her feet.

'Relax.' Emma loaded the trolley. 'Room service, for the dishes.'

'Wait.' Jaq put the chain on the door and called out. 'Who is it?'

No reply.

The rasp of a key in the lock.

Someone using the pass key. *Credo!* Jaq rammed a shoulder against the door and gestured to Emma to get out of sight.

The door moved; Jaq slammed it closed again. She put her back to the door, leaning against it, panting. A sudden force threw her across the floor. The door flew open as far as the chain would allow. Emma screamed.

Jaq caught sight of a man through the gap, his leg raised for another kick. Black beard and thick brows. Boris.

'Emma, call hotel security, the police . . . anyone!'

Boris kicked the door again, but the chain held.

Emma yelled into the phone as she ran a hand along the window, looking for a way to open it.

Boris raised a gun, pointing it at Emma silhouetted against the window.

'Get down!' Jaq hurled herself across the room and lunged at Emma, pulling her to the ground as two shots rang out.

'Stay down.' Jaq crawled away from Emma and grabbed the table, hurling it against the window. The glass disintegrated, showering onto the roof of the canopy below. Thank God for shoddy construction.

More shots, but the door held. Jaq pulled the quilt from the nearest bed and wrapped it around Emma. 'When I say go, you jump out of the window. Okay?'

Emma's face was as white as the quilt, but she nodded.

Kick. Kick. Smash, splinter. The chain was giving way.

Jaq crawled under the bed and across the floor. With her feet she opened the wardrobe. The trouser press fell out and she jammed it under the lever of the room's door handle, wedging it closed. Never had a trouser press been more useful.

Bullets were flying everywhere, deflected but not impeded by the door and trouser press. Jaq looked around for something else, anything she could jam against the door. Just to buy Emma a few minutes, long enough to get her friend out of here. At the noise of a police siren racing down the boulevard, the shots stopped.

'Go!' Jaq shouted.

Emma approached the window. She glanced over her shoulder, trembling, uncertain.

Something was happening on the other side of the door, the slap of plastic, rasp of a match, the crackle of a burning fuse and a faint but familiar scent of almonds. The trouble with Semtex is the smell . . .

'Jump!' Jaq yelled. 'Now!'

Emma screamed. The explosion lifted the door from its hinges, shattering it into smithereens, lumps of MDF and splinters of wood flying everywhere. The twisted metal frame of the trouser press hurtled across the room and smashed into the wall. Jaq was thrown to the ground, the signature blast ringing in her ears. Heart pounding, she waited for Boris to emerge through the dust, her fingers scrabbling among the splinters for a weapon to defend herself.

Was this it? Death in Belarus, the evidence destroyed by Semtex, her long and dangerous quest in vain?

A low groan. She crawled towards the noise. It came from the corridor. As the dust settled, she could make out a human shape on the other side of the demolished doorway. Boris lay on the blackened carpet, bleeding from a cut to his head. That would teach him not to play with explosives. But where was Emma?

Had she jumped? Had she fallen? Jaq struggled to the window. Emma lay motionless, lodged between the canopy and the wall of the hotel. A flower of red blood blossomed out through the white quilt that enveloped her.

'No!' Jaq hurled herself through the window. She bounced on the apex of the canopy and landed on the wrong side, rolling towards the street. Her fingers squealed against smooth polycarbonate as she scrabbled for something to arrest her fall, a handhold, anything.

She slipped over the side. In desperation, she grabbed at the guttering and hung on, her legs dangling over the pavement ten metres below.

The plastic guttering creaked, it was moving, parting company with the canopy.

How do you attach a uPVC half-pipe to a sheet of polycarbonate? With glue? With screws? What force will it withstand? A 62-kilo woman pulled by gravity at 9.81 metres per second squared gives a force of about 600 kilonewtons. Significantly more than the design case based on rainwater and leaves, with the odd pigeon or rat thrown in for safety.

The join began to weaken, to creak, to rip, to unzip.

Please! No!

As the plastic gutter gave way, Jaq fell onto the stone pavement below.

Saturday 9 July, Minsk, Belarus

Jaq opened her eyes and looked up to heaven. *Where was Emma? She mustn't be hurt. Emma the soft, the resilient, the unflappable, the indestructible – please God, don't let something have happened to Emma! Anything but that.*

A car screeched to a halt and the siren stopped. Doors opened.

'*Stop! Ruky vhoru!*'

Why were the policemen shouting at her? She wasn't going anywhere. Her hands were already thrown up behind her head as she lay spreadeagled on her back, unable – or unwilling – to move.

Boots thudded on the pavement beside her, then sirens, cars screeching to a halt, warning shouts, gunfire.

What to say to Johan, to little Ben, to baby Jade? How would they live without Emma? *Valha-me Deus.* How would she live with herself without Emma? It was all her fault. All those deaths. Sergei, Bill, Stefan, Petr. What had she done? What had she become? Please God, let Emma live. Take me, but save Emma.

Something clinked beside her. She turned her head to see shards of glass slithering over the edge of the canopy and shattering on the pavement. She needed to move. Why bother? What did it matter? What did anything matter if Emma had been shot?

A shower of glass was followed by a familiar voice.

'Jaq, are you okay?' Emma's freckled face peered over the edge of the canopy.

The surge of relief numbed all pain. Jaq shot up, reckless, and hooted with joy. *Merda.* What about Boris? She spun round in time to witness him being manhandled into a police van. *Graças a Deus.*

Only then did she check for damage. Starting with her hands, swivelling her wrists, bending her elbows, rolling her shoulders, rotating her neck, then lifting her hips, curving her knees and twisting her feet – no broken bones, only cuts and bruises. The gutter had ripped along its length, controlling her descent, slowing her fall. She'd taken a nasty blow to the back of the head. If her fingers hadn't felt the egg-sized lump pushing through her hair, she wouldn't have noticed it, running on adrenaline now.

A drop of blood fell onto the pavement. The fear returned.

'You're bleeding,' Jaq shouted up. 'Are you hurt?'

'Just a cut from the glass.' Emma held up an arm. 'I may sue the hotel.'

Another car pulled up. The door flew open and two Belarusian policemen leapt out followed by a sandy-haired man in plain clothes who ran towards her: Will-O'-the-Wisp.

Jaq met the clear green eyes of Detective Y'Ispe of the Slovenian Specials. She held up her wrists. 'Am I under arrest?'

'I have no jurisdiction in Belarus.' He flashed her a smile. 'You are not under arrest, but I could use your help.'

CODA

Thursday 14 July, Den Haag, Netherlands

The Organisation for the Prohibition of Chemical Weapons, OPCW, towers above the leafy centre of The Hague, hundreds of windows lit from within, unblinking eyes scrutinising the very worst of man's inhumanity to man. A Tower of Sauron searching for evil, the semicircular eight-storey building sits between Europol and the European Court of Justice.

Inside the operations centre, a bank of video screens streamed live footage from Chornobyl.

Detective Y'Ispe had downloaded the memory stick before driving Jaq and Emma back to the NATO base in Terespol, Poland. The Tyche tracker maps, together with Frank Good's belated testimony and Brigadier Dr Marion Fairman's resolute support, had been enough to convince the authorities to act. Jaq had begged to be allowed to accompany the inspectors, but the medical team were curt in their refusal; the brigadier offered to go herself.

The rapid response team were mobilised in record time, so that by the time Jaq arrived in The Hague, the expert operatives were already jumping out of helicopters over the Chornobyl complex.

Now Jaq was to take an adviser's seat in the OPCW control room.

'Welcome.' A grey-bearded man dressed in a dark suit extended a hand and introduced himself as the liaison officer. He hurried Jaq from the entrance steps to the operations centre. Three men and two women sat in front of a bank of screens. The liaison officer indicated a chair in front of the central screen and handed Jaq a headset.

Each screen showed the complex from a different angle. Separate teams dressed in hazmat gear – inflated silver spacesuits with self-contained breathing apparatus – moved stealthily towards the production building. The head and tail of each team gripped automatic weapons; those in between carried a variety of probes and portable analytical devices.

'Dr Silver has joined us,' the liaison officer announced.

One of the spacesuits stopped and saluted the camera. 'We found your factory, Dr Silver,' Brigadier Marion Fairman announced. 'Or what's left of it.'

The image became jerky as her team entered the first production hall. It took a moment for the picture to stabilise. Even then it was unrecognisable. Gone was the laboratory of pain, the giant flask, the tall glass column, the fat condenser. Gone was the toxic golden liquid cascading over a blue helix. All that remained was a glittering carpet of broken glass. No working equipment, and not a soul in sight.

The deployment of the rapid response team was fast. But not fast enough.

The debriefing was delivered from a hastily constructed chem hazard operations tent inside one of the empty storerooms of the Chornobyl complex.

Brigadier Fairman pulled off her helmet and addressed the screen. 'Dr Silver, do you recognise this?' She held up the Tardis bag. 'We found it in the laundry.'

'Yes, it's my bag.' Proof that she had been in the complex.

'You said that you saw a fully functional factory. When was that?'

Jaq closed her eyes as she reconstructed the timeline. So much had happened to her since escaping from the complex. 'June,' she said. Two months and a lifetime ago.

'Well, whoever took charge of the exit has done a thorough job. The large equipment is empty.' The screen flicked through a series of images: the tank farm, production halls and utility sections. 'Looks as if any smaller specialist equipment has been removed, and all the glassware destroyed.' A close-up of empty fume cupboards and bare benches. 'Everything was thoroughly cleaned first. We can find no residue in the production equipment.'

'No evidence?' The anxiety crept into Jaq's voice unbidden.

'Plenty of evidence, m'dear. A thorough job, but never thorough enough. They forgot the waste treatment.' The brigadier smiled. 'Everybody forgets the waste treatment. Toxic residues in the effluent pipes and unburnt documents in a bonfire.' The brigadier held up a sheaf of fire-curled papers in a gloved hand. 'Customer names, addresses, material data sheets, instructions for use. There is absolutely no doubt that this was a chemical weapons factory. We'll clean up these documents and scan them over.'

Jaq collapsed back into her chair. It was over. She wasn't insane. Finally vindicated, the tsunami of emotion made her feel dizzy. Weightless with relief, she resisted the surge of triumphal elation. No. It was never enough to be right; no one succeeds in isolation. She'd failed to convince the authorities in time, had been too slow to collect the evidence, taken too long. The bastards had escaped.

The liaison officer tapped the shoulder of the man next to him. 'Check the customer addresses against the Tyche tracker maps and the last debrief with Dr Hatton.'

She sat up straight. Wait.

'Dr Hatton?' Jaq's voice trembled. 'Dr Camilla Hatton?'

'Yes, our technical expert.'

'But she works for The Spider. She was there, in the Chornobyl complex. He called her his Swedish expert. She was in on it, she was working with SLYV.'

Greybeard tapped a button on the control panel. 'Permission to go to Level Two?' A disembodied voice came over the loudspeaker. 'Permission granted.'

Greybeard stood up and stretched. 'It's a beautiful day out there. Can I tempt you to a walk in the park?'

Jaq and Greybeard walked out into the sunshine in the direction of Scheveningen Woods.

'Why don't we start at the beginning?'

She adjusted her stride to match his.

'Some say Chernobyl was the tipping point. The Soviet rush for nuclear power, the disastrous planned economy, suddenly exposed for what it was. The Soviets needed money. The arms race was a drain. America and the Soviet Union signed an agreement to destroy their stocks of weapons.'

They stopped to let a nursery group pass in the other direction, babies in multi-seat strollers and a crocodile of toddlers in checked gingham pinafores, linked to each other by a continuous yellow ribbon.

Greybeard continued. 'Camilla Hatton was a scientist supporting the non-proliferation negotiators with technical information. Her PhD was in design of tracers, a smorgasbord of subtly different designer molecules, radionuclides and volatile chemicals present in minute quantities. Selected to be harmless, decaying or dispersing over time, but resistant to destruction by repacking or processing. She started Tyche, a company dedicated to tagging intermediates that might be used to make chemical weapons.'

A light breeze swished through the trees, bringing with it the smell of fresh pancakes from Madurodam, the miniature model village.

'Tyche invented a tracking device. Have you seen one of their trackers?'

Not just seen one, but searched for it, recovered it, unlocked it, marvelled at it and then lost it. 'Yes.'

'The Tyche tracker is a hybrid Geiger counter, chromatograph, spectrometer, GPS and information portal. It picks up the signals from detectors at border crossings, motorway tolls, bridges and ferries and converts the data into tables and maps. Everything you need to track a substance from source to destination.'

A family group were playing a noisy game of hide-and-seek

next to the lake. Giggles and whoops of excitement mixed with the quacking of ducks and rustling branches.

'The technical challenges were immense, but Camilla led a superb team. Tyche was a runaway success. All the Western chemical companies agreed to add tracers to the intermediates of interest. Using the Tyche tracker, we could see who was still active. And intervene.'

Bored with hide-and-seek, a couple of children had armed themselves with sticks and were engaging in a mock battle among purple azaleas. *Rat-a-tat-a-tat.*

'By 1998, eighty-seven countries had ratified the Chemical Weapons Convention. Tyche expanded in response, developing expertise in decommissioning. Collection and disposal was all done discreetly, safely and efficiently. Tyche won most of the government contracts.'

They left Westbroekpark and continued towards the sea.

'So, what happened?' Jaq asked.

'Pauk Polzin was appointed as finance director of Tyche.'

'The Spider.' Jaq shuddered.

'Business was changing. By 2005 Tyche had dealt with the known stocks of chemical weapons and the board claimed that new blood was needed.'

Greybeard stopped at a pedestrian crossing, waiting for the lights to change.

'Pauk Polzin had an agenda of his own. Instead of bringing the promised financial discipline, he opened the purse strings wide, allowed the scientists to try anything they wanted. They invested in new technology, bought new equipment, requested isotopes of ruthenium, lutetium and even francium.'

The wind became stronger as they approached the sea, bringing with it the familiar tangy smell of salt water and seaweed.

'A smokescreen for Pauk to channel money and materials elsewhere. When Camilla found out, she suspended him immediately and initiated criminal charges, but the board overruled her, decided to hush it up. They cut a deal with him to keep quiet. No one realised how bad things were. Pauk brought the company to the brink of ruin. By the time he left, Tyche was already so heavily in debt to Zagrovyl that it was technically insolvent.'

A train rattled past.

Things began to fit into place. Tyche bought francium from Zagrovyl. The price was ruinously high, the profit margin astronomical. Frank named a yacht after his windfall from the deal.

'National governments had invested a lot of money. When Zagrovyl offered to buy the company and write off all the debts, the board jumped at the chance. Tyche was quietly subsumed into Zagrovyl.'

'And Camilla?' Jaq asked.

'Zagrovyl gave her an empty role until she resigned.'

So, she wasn't lying. She did work for Zagrovyl, albeit briefly. Director of Change. Was there ever a company less willing to change?

They crossed the road to the beach, a vast expanse of golden sand disappearing into the pale grey North Sea.

'But she was there, with The Spider. How could she have betrayed everything she had worked for?'

'Camilla was working undercover for OPCW.'

A wave crashed onto the beach. A sudden slap – worse than a slap – a 1000v electric shock administered through a shower of icy water.

'You mean, you knew about the complex? All this time?'

The anger began to build. They knew, and yet they let the police arrest her, allowed her to race across Europe on a wild goose chase, left her to the mercy of The Spider and SLYV.

'We couldn't have found the complex without you.' Greybeard put out a hand as if to touch her, withdrawing it as she flinched. 'We suspected its existence, but we hadn't been able to pinpoint the location. Could have been Ukraine, Belarus, Moldova or Russia. All the tracking signals went cold near the Chornobyl zone of alienation. Without a definite location, without concrete evidence, we couldn't get international permission to go in. Russia is particularly sensitive about what they see as Western interference in Ukraine. Dr Hatton tried to use stronger tracers, developed a more sophisticated tracker to obtain proof, but something went wrong.'

Jaq's anger evaporated as quickly as it rose. It was her fault that something went wrong. Her interference that had messed up the tracking operation.

'Camilla decided to go undercover, offer her services to SLYV. We advised her against it, told her keep away from Pauk Polzin, all too dangerous, but she insisted.'

'So where is she now?' Jaq looked around, expecting Camilla to emerge from the sea in a turquoise wetsuit.

'I'm sorry, Jaq, I thought you knew.'

'Knew what?'

'She didn't make it out.' Greybeard dropped his eyes. 'Camilla Hatton is dead.'

Friday 5 August, Melrose, Scotland

Jaq followed the path beside the river until she judged it level with the monastery. The low roar of peaty-brown water, tumbling and foaming, accompanied by a descant of birdsong, drowned out any human noise. Ozone, terpenes, esters – the complex natural scents of the forest – wafted past on a light breeze. She turned away from the river and struck a perpendicular path, climbing steeply through a forest of beech trees. A canopy of lime-green and copper leaves sheltered her from the pitter-patter of light summer rain. A grazing fawn froze at her approach, dark eyes staring deep into her soul, before he skittered away over a carpet of moss. Away from the river, close to the brow of the hill, the music was unmistakable: Bach's Toccata in D minor. She had absolutely no doubt who was playing the chapel organ.

As she breasted the hill, the complex lay before her in a secluded valley surrounded by wooded hills. The evening sun dipped under the rain clouds and lit up the stone buildings, rose and gold, nestling on the site of an ancient Cistercian monastery. The super-rich paid extra for bare cells, plain food and total isolation. No internet, no TV, no phones, no press and enhanced security.

Jaq followed the electric fence until she found a junction box. Pulling the tools from her backpack, she worked swiftly. Once the power was off, it took seconds to scale the fence and jump the barbed-wire ditch. She crouched low and kept to the shadows until she reached the chapel door.

The cast-iron latch was cold to the touch. It slid back smoothly, and she pushed at the heavy oak door. The creak was drowned out by a furious four-part fugue.

He had his back to her, white shirtsleeves billowing as his hands moved across the multiple keyboards. His fair hair was longer, curling over his shirt collar to the grey silk of his waistcoat. She waited until the final coda before approaching.

'Hello, Frank.'

His eyes darted up to the mirror above the keyboards – an organist needed to watch his choir – hard blue eyes that displayed no surprise.

'Jaqueline.' Frank remained seated. 'An unexpected pleasure.'

'I'll take that as overdue thanks for saving your worthless hide.'

'You can take me any way you want.'

He gestured to a chair beside the organ stool. She didn't move.

'Not a social call, then?' He leafed through the musical score. 'I thought not. So, what brings you here?' He began to work the foot pedals again.

She leant over and closed the valve for the air pump. The hissing stopped.

'Pauk Polzin,' she said.

His eyes glinted in the mirror. For the first time she caught a glimpse of the fire under his icy carapace. There was a man underneath the mask. An angry man.

'The Spider,' he hissed. 'That double-crossing bastard.'

'Pot calling the kettle black?' she said. 'You are a fine one to talk.'

This time his smile was almost human.

'Touché, Jaqueline. But I think you will agree, the circumstances were somewhat extreme.'

'Business as usual for you, I would have thought, Frank. Stomping on the necks of others. Isn't that how you crawled your way to power?'

'Power,' he said. 'An interesting concept. And as it transpires, fairly meaningless when you are locked in a deep freeze.'

'I should have left you there,' she said.

'Yes,' he said. 'In your place I would have done exactly that.'

He opened a new score and began adjusting the stops on the

console, pushing the ivory circles back to their default positions, resetting the instrument.

Cabrão. 'What did you do with the Tyche tracker?'

'Alas.' He shrugged. 'Destroyed by The Spider.'

Liar. 'How did you survive the crash?'

'Lucky, I guess. Flung out. Some peasant found me in a bog. I offered him money to alert the nearest Zagrovyl outlet. Or so I'm told. I still don't remember much.'

She watched him as he shuffled into position. Preparing to play as if nothing had happened.

She tugged at the heavy wooden lid. He pulled his hands away just before it smashed down, covering the upper keyboard.

At last she had his full attention. He twisted round to face her. 'Turns out we're not so different, you and I.'

'If you exclude decency, morality, honour, empathy . . .'

'We are survivors, Jaqueline. The qualities you describe are simply adaptations of the weak.'

She brought her face close to his.

'Why didn't you back me up? Why didn't you go straight to the police?'

He shrugged. 'I made a full statement.'

'Only after Johan—'

'Your mountaineering friend? Yes, I am most grateful to him for his shock treatment. It cured my temporary amnesia.' His eyes narrowed. 'A fine specimen of a man. Another lover? You seem to favour the rugged outdoor type.'

Jaq clenched her teeth. She wasn't going to rise to the bait.

'You only told part of the story.'

'I told the police everything that mattered. Thanks to me, a dangerous illegal factory has been closed.'

The arrogant, insufferable prick. 'We only recovered part of the data.'

Frank nodded. 'The important part. Enough to convict the producers and track down buyers of chemical weapons.'

It was true that there was more than enough evidence to make the arrests. Dawn raids on a beauty spa in Luxembourg, a pet food distributor in Ireland, a currency trader in Switzerland, an organic food wholesaler in Andorra, a specialist travel company in Pakistan. But that was not enough.

'And what about the suppliers?' she asked. 'What about Zagrovyl's part in all of this?'

'Zagrovyl played no part. Except as an innocent victim, caught up in The Spider's web.'

Was he telling the truth? Did it matter any more? If nothing else, he was guilty by omission.

'Because of you, Camilla is dead.'

His mouth tightened. 'I'm sorry for your loss.'

'Because of you, Pauk had time to get away.'

'I'd be more than glad to rectify that.'

Time to get what she came for. 'You have a yacht.'

'I do. Berthed in Cannes. Named after the element that made my fortune.'

Francium. A half-life of twenty-two minutes. Never more than an ounce – a few grams – in existence at any one time in the whole of the earth's crust. The jewel in the crown of Zagrovyl's precious metals empire.

'Tell me, what is your interest in *Good Ship Frankium*?'

'Francium.' Jaq corrected him. The ephemeral element discovered by a French chemist who named it after her country. 'Soft for France, not hard for Frank.'

A flash of irritation erupted through his thin veneer of civility. 'It is my yacht, and I call it what I like.'

'I need it.' There, she had said it.

'Indeed.' Frank raised an eyebrow, his vile snake-like smile back now that he knew she wanted something from him. 'It's not for sale. And in any case,' he looked her up and down, taking in the black leggings, cheap fleece and muddy boots, 'I doubt you could afford it.'

'I'm going after Pauk.'

Frank stood and paced over to the stained-glass window. The evening sun threw coloured light onto his pale skin in ruby, emerald and turquoise patches. He rubbed his chin with the back of a hand.

'You know where he is?'

'Maybe.'

'But you don't trust me enough to tell me?' Frank nodded. 'Fair enough.' He drummed his fingernails against the lead moulding. 'You need a yacht, so he must be by the sea.'

Jaq said nothing.

'Now you come to mention it, I think Pauk did have a seaside villa. Somewhere hot. Thalassotherapy for his chronic pain. Caribbean? Mediterranean? No, Interpol could get him there and you wouldn't be bothering me. So, somewhere without an extradition treaty. The other side of that rusty old Iron Curtain. On the Black Sea? Crimea, perhaps?'

Jaq remained silent.

'Yes, I see now. Hypothetically speaking,' he said. 'If I were to lend you my beautiful yacht, would that make us even?'

'Not even close,' Jaq said.

Frank twisted his mouth into a rueful smile. 'I thought you might say that.'

This was a mistake. 'Forget it, I shouldn't have bothered coming.' Jaq spun on her heel and strode towards the door.

'The answer is yes.'

Jaq stopped. 'Yes to what?'

Frank returned to his stool. Flinging the lid open, he set up his music above the keyboard before flicking the air switch. The hissing resumed, snaking round the little chapel.

'You can take my yacht to go after The Spider. I ask only one thing in return.'

'You are in no position to—'

'Bring him back alive,' Frank snarled. 'We have unfinished business.'

Thursday 1 September, Crimea, Ukraine

Waves foamed against the sleek composite hull and the laminate sails flapped as Jaq gybed and luffed onto the new course. The water shimmered blue-green and inviting in the early-morning sunlight, the Black Sea far from black today. The Crimean cliffs were almost visible with the naked eye. Getting closer. Binocular distance. Time to conceal identity and purpose.

She had peeled the misspelt name of the yacht from the hull before they set sail, and now she let the halyard fly. The ripstop-nylon spinnaker pooled and slithered across the polished wooden deck, concealing the giveaway elemental symbol emblazoned in orange across the billowing yellow: 87 – Fr – 223.

Francium. An element waiting to be found for hundreds of years. A hole in the periodic table, the missing alkali metal, claimed first by the Soviets as russium, then by English chemists as alkalinium, by the Americans as virginium, by a Romanian as moldavium and by others as eka-caesium. They were all trumped by French scientist Marguerite Perey, who finally proved its existence beyond doubt: francium, the ephemeral element.

Giovanni, the skipper, emerged from the galley. Dressed in crisp white cotton shorts and a short-sleeved dark blue linen shirt, he hummed as he set out her breakfast. Perfectly poached eggs, thick hollandaise sauce and emerald-green spinach on proper English muffins. Nice.

He inspected the spinnaker bag. 'Good work.' He took the helm while she ate. 'You'll put me out of a job soon.'

Jaq smiled and waved her fork. 'I'll be captain if you'll stay on as cook.'

'I could live with that.' He appraised her, eyes narrowed, lips slightly parted, white teeth gleaming.

Jaq turned away to hide her hunger, a warm flush rising through her body. Behave. Not now. Work to do. A Spider to snare.

The yacht rounded the westerly cliff, and Giovanni dropped anchor. When the sun was directly overhead, Jaq stripped off and dived over the side of the yacht. She rounded the promontory and swam freestyle, long, smooth strokes, parallel to the shore. Stopping to get her bearings, she lay on her back and let the sea swell lift her up and down as she surveyed the Crimean Riviera. Mountains rose sharply from the sea, white limestone sparkling in the sunlight, winding roads visible from the lines of tall dark green cypress trees. On top of the promontory, the Swallows' Nest dominated the skyline, a fairy-tale clifftop castle, a twentieth-century folly, a temple to oil money.

Gliding through the water, cool silk on her skin, Jaq changed to a smooth underwater breaststroke as she stole into the hidden cove. Holding her breath, she located the underwater jetty, the mooring rings shiny with recent use. She came up for breath and trod water until she located the villa, hidden behind mature umbrella pines.

Giovanni was waiting, pacing, his plimsolls squeaking on the wooden deck. He extended a hand as she mounted the polished steel ladder, his palm hot against her sea-cold wrist. '*Che gelida manina*,' he sang.

A boy who knew his Puccini. Nice voice. *Se la lasci riscaldar?* Later. She grinned at him. 'I found it.'

Giovanni handed her a towel. 'I never doubted you.'

She dried her hair before wrapping the towel around her hips. 'The Spider's lair,' Jaq said. 'A deep cove and an underwater pier. Perfect.'

Riccardo and Lorenzo reported for duty.

Jaq surveyed her crew. A hand-picked team. Johan had recommended Giovanni – a climbing buddy who skippered yachts for a living – and Giovanni had found the muscle: former GOI

– Gruppo Operativo Incursori – elite commando frogmen of the Italian navy, now private military contractors. Men for hire. But this time, she was calling the shots. She unpacked the explosives and gave them instructions.

Interpol couldn't risk anything official, wary of upsetting Russia or the Ukraine, but they were willing to turn a blind eye to Jaq's plan. If Pauk were to board the boat willingly, Lorenzo's men could arrest him twelve nautical miles from the shore. Then the *Frankium* would cross the Black Sea, sail into the Mediterranean and return The Spider to the rule of European law.

European law. A curate's egg. Good in parts. The autopsy on Stefan Resnik showed no traces of ergotamine in his blood. Just alcohol and paracetamol. A review of the medical evidence concluded that he had suffered a heart attack while being suffocated – presumably with his own cushion, since fibres matching the one on his chair were found in his lungs. Poor Stefan. The forensic evidence from the investigation into Sergei Koval's death included a set of keys, still intact after the explosion. Sergei's keys. The tin under a drain cover used to conceal the keys was discovered and checked for fingerprints. One set matched those in Stefan's flat. They belonged to a man held in custody by the Belarusian police, a man who had been interviewed by the Cumbrian police shortly after William Sharp's death: Boris Cimrman. All charges against Jaq were dropped.

Time to bring the ringleaders to justice. The fact that Frank Good was funding her to collect the evidence against him made the voyage all the sweeter.

'Get some rest,' Jaq said. 'We go in tonight.'

Jaq surveyed the terrain around Pauk's villa. First from the sea. Her triceps and quadriceps were burning as she paddled the inflatable kayak, hauling it round the western headland, keeping her distance from the underwater jetty. Johan would have laughed at the clumsy craft – the blunt profile made it sluggish in the water – but for this mission stability and storage space were more important than speed. The villa remained hidden, the only clue to human habitation the noises coming from an old boathouse, the rattle and chug of a diesel generator. Judging by the continuous wisps of steam, Pauk was a heavy power user.

Ideal.

Jaq wrestled the kayak past the eastern promontory and came ashore in the next bay. She lugged it from the water and opened the air valve, hiding the deflated boat and its contents under branches and leaves. She laced up her hiking boots, slung a coil of rope across her chest and began her exploration on land.

A cliff rose steeply from the forest: Jurassic limestone dotted with hornbeam, oak, maple and pine on the lower slopes, rising up to the bare, jagged white peak of Roman-Kosh, the highest mountain in the Yayla range. Pine needles snapped underfoot, and the burnt sugar scent of succulents rose from the ground as she climbed towards the road snaking up from an eastern port to the Crimean plateau. When she calculated that she was directly over Pauk's villa, she scrambled down through flowering bushes and tangled creepers to the precipice. She lay down and edged forward. The agglomerate crumbled and little pebbles of friable karst tumbled towards the forest below. She slithered backwards. The villa was invisible from above, and the overhang was fragile.

Perfect.

She returned to the road and continued until she stood on the other side of Pauk's bay. Somewhere here lay an outcrop of granite. Thank goodness for geological maps, for scientists, the patient factfinders and rigorous recorders of the natural world. A change in vegetation, as the soil became acidic, was all the clue she needed. The spine of hard rock was hidden under lush evergreen vegetation. Jaq tied the rope to an umbrella pine and abseiled down, towards the yacht moored in the western bay, avoiding the crumbly limestone on either side, keeping low until she had a viewpoint.

The villa was completely hidden from the road, but hanging from a rope halfway between cliff and sea, Jaq had a bird's-eye view. Constructed on three levels, the building was cleverly disguised to blend in with the forest, the glass windows of the top floor coated in a tinted film to avoid any glint from reflected sun. The windows of the lower floors were obscured by trees. This was someone who valued their privacy more than sea views.

Three ways in and out. Air. Land. Sea.

The main door was on the middle floor, a massive construction of oak and steel. A paved path led under a natural limestone arch into a huge sinkhole, where an ancient stream had pooled, dissolving the rock to create a sheltered helicopter landing pad at the west side of the villa. The chopper must come in from the sea and drop before hitting the cliff. It would require a skilled pilot to land in the sinkhole. Someone like Mario. On call 24/7. Escape by air.

The helicopter garage was empty except for cigar butts. Definitely Mario.

The second door led from the top floor, over a balcony, across the roof to a steep path, stone steps cut into the cliff behind. A heavy iron gate set deep into the rock marked the boundary, a narrow culvert heading to the road. An old stream bed, no one viewing it from above would guess it connected a villa to the road. Escape by land.

The third exit was from the basement, a tunnel leading from the side of the house to the boathouse and underwater jetty. Escape by sea.

Form follows function. The plan was taking shape.

Jaq climbed back up the granite spine to the road. A warm tingle of anticipation quickened her movements as she hid the rope and dashed through the forest to the beach. She whistled to herself as she unpacked the explosives from the deflated kayak and got to work.

Not long now.

Giovanni made the broadcast at dusk, hacking into the satellite feed with pre-recorded images of a terrorist attack in Sevastopol and Yalta. The Crimean Liberation Front were shown calling their fellow Ukrainians to arms. Angry over the increasing Russian presence after the Kharkiv Accords, they called for closer ties with Europe, membership of NATO and a stop to the creeping Soviet presence in Crimea. The film cut real footage of Ukrainian politicians with stock scenes of marauding gangs and ended with a studio warning to all Russians to seek places of safety.

Thirty minutes later the helicopter appeared, a dark, sinister insect flying across the bay without lights. The pulsating orange circle of a cigar end lit the pilot's features as it banked overhead.

The man who'd murdered Petr in cold blood: Mario.

Riccardo sailed into the deep cove in front of Pauk's villa and dropped anchor.

Hidden in the woods, Jaq waited for the helicopter to land. Then she counted down.

One.

Two.

Three.

This one is for you, Petr.

She pressed the detonator.

The first explosion, a blue flash, severed the main cable from the generator, plunging the villa into darkness.

The second explosion, seconds later, a thunderclap under the landing circle, destroyed the helicopter, smashing the fuel tank, igniting the heptane. Tongues of flame snickered into the dark sky.

Lorenzo was right on cue, bellowing a warning from a foghorn overhead. Announcing the attack of The Crimean Liberation Front. Death to all Russians.

Now Pauk would know the helicopter had not crashed by accident; the blackout was not a mechanical fault. He was under attack.

The third explosion destroyed the cliff overhanging the villa.

A muffled blast then a sigh as hundreds of tonnes of limestone hurtled down the cliff, smashing through trees, bouncing and rolling and clattering onto the roof of the villa. A distraction to allow Lorenzo and his team to abseil down and open fire.

Bang, crack, whizz. More explosions, irregular, seconds apart, made it sound as if Lorenzo had an army with him. Never had firecrackers in trees been so effective.

Pauk had only one escape route left. Towards the sea.

The GOI medic manned the lifeboat. He switched on the searchlight and swung it towards the shore.

A tall, thin man loped towards the jetty, followed by four armed thugs. He waved at the boat and shielded his eyes.

'Rescue,' Riccardo yelled from the yacht. 'Quick, get on board.'

One of the bodyguards raised a gun, taking aim at the medic in the lifeboat. Lorenzo was too quick for him. Darting out from the trees, he shot the thug through the head. Pauk screamed and ran into the sea, leaving his bodyguards to fight it out on the beach.

'Get me out of here,' Pauk demanded.

The medic hauled him into the lifeboat and sped back out to sea.

As the lifeboat approached the yacht, Riccardo threw down a ladder. He held out a hand for Pauk.

'Welcome aboard.'

Thursday 1 September, Crimea, Ukraine

Jaq surveyed the remains of the helicopter, a tangled mass of metal and glass. Somewhere in there lay the charred corpse of the pilot, cigar-smoking Mario. She bowed her head. Where was the elation? For all his crimes, he was a human being. A human being she had murdered. It wouldn't bring Petr back. She turned away.

'Come quick,' Lorenzo shouted. 'There's a prisoner in the cellar.' He radioed for the medic and dashed back into the building.

Jaq paused at the top of the stairs. Below her, a woman sat cross-legged on the concrete floor, dressed in an orange jumpsuit, a heavy chain extending from rings round her ankles, through manacles on her wrists, to an iron ring in the wall. White hair, skin pale and drawn, the heavy black shadows framing green eyes which flashed as she raised her head. Jaq's heart soared.

'I should have guessed it was you.' Camilla's voice had lost none of its mesmerising gravitas. 'You do have a rather distinctive taste in explosives.' She held out her wrists for Lorenzo to unlock the irons. 'Is it safe?'

'Five X-rays neutralised, ma'am,' Lorenzo announced. He beamed at her, proud rather than remorseful: the right man for the job. 'Target secure in the bilges.'

Jaq ran down the stairs and pulled Camilla into an embrace. The older woman trembled in her arms, lighter and frailer but very much alive.

'Nice boat.' Camilla raised her coffee cup in a toast. After a thorough medical she had been allowed to wash and rest, and now she looked almost restored. The twinkle was back in her eyes.

'Simplest way to cross borders with explosives.' Jaq refilled the coffee cups from a cafetière as Giovanni removed their breakfast dishes. 'How long—'

'In Pauk's dungeon?' Camilla shuddered and sank back into her seat. 'Since the visit to the complex in June. Turns out I'm not very good at lying.'

'Are you . . . okay?'

'We go back a long way, Pauk and I. He thought he could extract information. But I wouldn't give the bastard anything.'

'Everyone thought you were dead.'

'I would have been if you hadn't come. Pauk was running out of patience.' Camilla paused. 'Did you find Sergei?'

'I'm sorry.' Jaq held her gaze. Brown on green. 'Sergei died in the explosion at Snow Science.'

Tears formed in Camilla's eyes, spilled over and ran down her cheeks. Jaq sat beside her in silence, listening to the waves lapping against the hull while the woman wept.

'He was a fine man.' Camilla wiped her eyes. She ran a hand through hair that had grown long in captivity, smoothing it away from her brow. 'The only helicopter pilot who managed to follow the tracer signals all the way to Chornobyl. Where they vanished.'

'Was that when Sergei vanished, too?'

'No.' Camilla stared out at the sea, her voice low and desolate as she continued. 'Sergei left Snow Science on my instruction, to plant a special tracker in the zone of alienation.'

'Why didn't you tell me?'

'I didn't trust you.' Camilla turned away. 'I forced Laurent to destroy the samples, threatened to pull his Zagrovyl funding if he didn't comply.'

'It was you!'

'I needed the delivery to complete. I had to prove what was happening, trace it all the way. It wasn't enough to pinpoint the weapons complex, I needed to track right to the end customers. Then, and only then, would there be enough proof.'

'It worked, Camilla. Your tracer and tracker worked. OPCW got the evidence they needed.'

'*Tak Gud*.'

Jaq stood and walked to the opposite window. The sky had darkened. Heavy weather ahead. 'There's something I have to ask you.' She inhaled deeply and let her breath out slowly. 'Did you pay Karel?'

'Who?'

Jaq swallowed hard. 'The ski instructor in Kranjskabel.'

'Oh, you mean Sergei's skiing friend.'

Jaq spun round, eyes flashing as the last piece of the puzzle fell into place. Karel attended Sergei's funeral. She hit her brow with the flat of a hand. 'Karel Žižek was a friend of Sergei Koval?'

'Yes,' Camilla said. 'He asked Karel to keep an eye on whoever took over from him at Snow Science, lent him his flat.' She smiled. 'I don't imagine he expected Karel to get so deeply involved. By the look of things, it was no hardship to either of you.'

Jaq frowned. 'But the ergotamine?'

'It was clumsy, I know, but I had to protect you from yourself.'

'It was *you*?' Not Karel.

'They killed Stefan, they were coming after you next. Denouncing you for Stefan's murder was expedient – it forced the police to act fast. I planted the ergotamine later, when you were in custody. The poison is connected to you through your scientific paper, but it's easy to analyse for. Once they had Stefan's toxicology the case would be thrown out and you would be released. I just wanted to make sure you were safe for a while. But once again, you took matters into your own hands.'

Jaq bit her lower lip. Had she misjudged Karel? She'd escaped because she had no choice; leaving him to face the police, the courts, perhaps even to be charged as an accomplice to her flight, by omission if not by action. The affair was over, nothing would change that, but she owed him an explanation.

Jaq returned to the seat opposite Camilla. 'What about Frank Good?'

'What about him?'

'He must have known what was going on.'

'I don't think so.' Camilla held up a hand to interrupt her protest. 'Frank hired Pauk to investigate the misdirection of Zagrovyl materials destined for Smolensk. Once he realised what Pauk was up to, he refused to deal with him. Frank is a poor judge of character – he picked the person masterminding the scam as investigator – but I don't believe Frank was involved in supplying chemical weapons.'

Jaq frowned. 'Then why did Frank steal the Tyche tracker from me?'

'Pauk told him you had information incriminating Zagrovyl. He was right.'

'Frank denied you were ever a Zagrovyl employee.'

'At my request,' Camilla said. 'I resigned the day after I met you. I asked Zagrovyl to destroy all trace of me. Frank got a private investigator to wipe my record clean.'

Bill Sharp. The man who drowned in Lake Coniston. Mr Beige. Murdered by Boris.

'I didn't want to be associated with Zagrovyl and, more importantly, I had to stay under the radar to collect the evidence to catch and convict Pauk. Frank can be useful at times.'

Jaq shook her head.

'Frank needs to pay for his actions.'

'He may not be a saint, but he respects the rule of law, at least insofar as it suits his self-interest.'

'Frank lied.' Jaq thumped a fist down on the bench. The padded leather sighed and wheezed as it resumed its shape. 'He claimed he had never been in the complex, denied my role in his escape.'

'Until you and your friends persuaded him otherwise,' Camilla said. 'And then he came clean.'

Frank's statement to the police had been a masterpiece in self-aggrandisement. He had managed to confirm the existence of the SLYV complex without in any way incriminating himself. Painting a picture of a nerve-agent-damaged, amnesiac hero.

'Only to confirm the complex existed. And by then it was too late.'

'Not at all,' Camilla said. 'Pauk did the dirty work of decommissioning for you. He is nothing if not efficient.'

'How do we know SLYV won't set up a chemical weapons factory elsewhere?'

A thud and cry below deck. Lorenzo and Riccardo interrogating Pauk in the prison cell they'd constructed in the bilges.

'Because Pauk Polzin is going to be locked up for a very long time. Without The Spider, SLYV is nothing.'

Jaq paced the smooth wooden floor. 'It doesn't feel right to let Frank get away scot-free,' she said.

'If we expose Frank, Zagrovyl will sack him.'

'Good,' Jaq said.

'Not necessarily. Someone else will take his place. Someone equally ruthless and rotten.' Camilla sat forward. 'The problem is not Frank per se, the problem is the way people are promoted and rewarded in companies like Zagrovyl. So long as boards try to manage complex organisations using simplified incentives, it's the ruthless, immoral Frank Goods of the world who rise to the top.'

'Maybe I can't fix the system,' Jaq said. 'But I can ensure Frank is punished.'

Camilla shook her head. 'Pause for a minute. Think how useful he could be to us. He knows we can destroy him. If he continues in post, we can influence him, we can be his conscience.'

'So what are you proposing?' Jaq asked.

'I'm proposing we do nothing.'

'Nothing?'

'For now.' Camilla smiled.

Inaction? Unthinkable. Infuriating. Even more so because

Camilla was right. There were plenty of greedy, soulless Frank clones in Zagrovyl, jostling for power, barking and howling to be top dog. Better the devil you know? She didn't like it. Her sense of natural justice revolted, but she could see the logic.

'For now,' Jaq said. 'And then?'

Giovanni ducked his head through the door.

'Storm approaching,' he said. 'Last chance of fresh air before we batten down the hatches.' He held out a hand for Camilla, but his smile was only for Jaq.

The wind filled the sails and brought the smell of the sea. The salts of iodine and potassium, manganese and sodium have no odour. The scent comes from more complex chemicals – dictyopterenes, bromophenols, dimethylsulphide – the smell of seaweed and algae, the smell of nature.

The sea surrounded them, darker now, turbid and restless.

'Look.' Giovanni pointed. A pod of bottlenose dolphins leapt in and out of the bow wave, their sleek grey bodies soaring and diving in perfect arcs.

Giovanni stood behind Jaq, not touching. The heat from his body warmed her back – low-frequency, long-wavelength radiation because it passed through the skin, gently melting her from the inside out. Her breath quickened, matching the rhythm of his hot breath on the back of her neck. She tipped her nose to catch the scent of him, sandalwood and leather. And musk. God, it had been a while.

Camilla stood beside them, transfixed by the dolphins. As the boat pulled away and the sea grew dark, she yawned. 'Maybe I will rest for a while.'

Camilla stumbled as she turned. Giovanni leapt forward, offering his arm. 'Let me check you have everything you need.' He cast a long look back. 'Wait for me,' he mouthed before disappearing down into the galley.

Jaq slipped off her shoes and climbed the mast. She clipped onto the crow's nest rail and stared out to sea. The yacht was making good progress: the Crimean cliffs a glimmer to the north; land already visible to the west. They would enter the Bosporus by nightfall, spend tomorrow crossing the Sea of Marmara, through the Dardanelles and into the Aegean Sea.

Once she handed Pauk over to the authorities, there were things to do. Visit Cecile and baby Lily. Call Karel with an apology. Treat Emma and Johan and the brood to a big celebration. Resign formally from Snow Science. And then what?

With the loose ends tied up, time for a break. Time for herself. Sail the Mediterranean. Write a textbook on the safe handling of explosives.

Giovanni emerged from the galley. He stood on the deck, legs apart, one hand raised to shield his dark eyes from the sun as he stared up at her. He had a special sort of stillness, total control of every muscle. Frozen. Waiting to see what she would do. Anticipating her next move. The waiting game. Predator and prey.

Good things come to those who wait. Even better things come when you take control. The sails luffed in the strengthening wind, trembling, whistling: the call of the hunt.

Sparks flew as Jaq slid down the mast towards Giovanni.

Game on.

Author's Note

The Chemical Detective is a work of fiction, a story; I made things up.

There are, however, some real places and events – and some real science and engineering – blended in with the fantasy, so to clarify:

On 26 April 1986 an explosion at the Chernobyl nuclear power plant led to a massive release of radioactive material.[1] Thirty-one people lost their lives as a direct result, and many more suffered chronic ill health and premature death in the decades afterwards. The worst-affected area of Ukraine and Belarus – the zone of alienation – will continue to be unfit for human habitation for centuries.

The Chernobyl Shelter Fund is an international project started in 1997 to contain the crumbling reactor sarcophagus so that the remaining radioactive material can be safely removed.[2] In October 2017, the New Safe Confinement (NSC) was wheeled over the reactor at an estimated cost of 1.5 billion euros.

I was moved to tears by Svetlana Alexievich's masterwork *Voices from Chernobyl*, collected oral testimony from survivors. For anyone interested in more technical details, Grigory Medvedev's *Chernobyl Notebook* is a first-rate first-hand account. For those wishing to take a tour without leaving the sofa, Bill Murray's excellent and concise *Visiting Chernobyl* gives a lucid account.

1 About six tonnes (or 3.5 per cent) of the nuclear fuel was released to the environment, including radioactive krypton, xenon, iodine, caesium, strontium, tellurium, zirconium, niobium, lanthanum, cerium, neptunium and plutonium.

2 Of the original nuclear fuel (180 tonnes), about 95 per cent remains under the sarcophagus in the form of core fragments, dust and lava-like 'corium'.

The shadowy organisation SLYV, and the secret chemical weapons complex hiding in the shadow of the real NSC, are pure fiction.

Geiger counters and other scanners are used to screen goods crossing borders and ports. Manufacture and trade in scheduled chemicals (those that can be used for both legitimate and nefarious purposes) is restricted under the Chemical Weapons Convention. Tracers are added to explosives like Semtex; some tracers have a distinctive scent.

The company Tyche, and the Tyche tracker, are fictional.

A hole in the periodic table[3] was filled with discovery of the element with atomic number[4] 87 (between radon, 86 and radium, 88). Claimed by many (with names russium, alkalinium, viginium, moldavium, eka-caesium, actinium K), the first confirmed discovery was by Marguerite Perey in 1939, who named it after her country, France.

Francium is an extremely rare, highly radioactive metal. The most expensive naturally occurring element in the world is found in trace quantities in uranium and thorium ores where it is continually forming and decaying into astatine, radium and radon. There are 34 known isotopes,[5] the most stable of which is francium 223 with a half-life[6] of 22 minutes.

Quite apart from the rarity and radiotoxicity of Francium (poor Marguerite died of cancer, like so many researchers from the Marie Curie Institute) it would make no sense to use Francium as a tracer, which is why fictitious company Zagrovyl discover a fictitious isotope of Francium with

3 Periodic table: a table of the 118 chemical elements arranged in order of atomic number, usually in rows, so that elements with similar atomic structure (and hence similar chemical properties) appear in vertical columns.

4 Atomic number: number of protons in the nucleus.

5 Isotope: variants of a chemical element with the same atomic number (same number of protons in the nucleus) but with a different number of neutrons and therefore a different atomic mass.

6 Half-life: the time taken for half of the material to decay.

fictitious properties that perfectly fit the bill for fictitious company Tyche who purchase it at ruinous expense.

ESA (European Space Agency), Earthwatch (connecting the general public with scientists to collaborate on environmental field research) and OPCW (Organisation for the Prohibition of Chemical Weapons) are all real organisations doing important work. OPCW won the Nobel Peace Prize in 2013 in recognition of its efforts to eliminate chemical weapons.

In its heyday, Imperial Chemical Industries (ICI) was the largest manufacturer in the UK. Formed in 1926, it had divisions manufacturing everything from explosives to fertiliser, pesticides to plastic, paint to precious metals, radioisotope trackers to fragrances. The main UK plants included Billingham and Wilton on Teesside. ICI sold off most of its business in the late 1990s and early 2000s in an unsuccessful attempt to improve profits by moving away from bulk chemicals. The company had all but completely disappeared by 2008, but its legacy lives on in Teesside with many of the old ICI factories still in production though operated by other multinational chemical companies.

Gaia Vince's brilliant book *Adventures in the Anthropocene*, with its fascinating story of artificial glaciers in Ladakh, inspired the idea of an alpine research organisation dedicated to snow and ice.

Snow Science is a fictitious research institute. Kranjskabel is a fictitious place, roughly located near the beautiful town of Kranjska Gora in Slovenia, but looking a lot like Meribel in France.

'Monday morning sudden death' was a recognised phenomenon among workers in dynamite factories, caused by a physical response to withdrawal from chemicals over the weekend. Nitroglycerine and nitroglycol cause blood vessels to widen. When exposure is interrupted, the blood vessels constrict, and in extreme cases, the reduced blood flow to the heart can lead to sudden heart attack and death.

There are no recorded cases of sudden death due to improved industrial hygiene. There is no nitroglycerine factory in Seal Sands.

And, just in case it needs to be said, although I may have stolen some names, none of the characters in the book represents a real person.

Acknowledgements

I started writing the Jaq Silver series of stories after a skiing accident gave me some unexpected time off work. Alas, my enthusiasm for the project vastly outstripped my skill. My best friend, Marjory Flynn, suggested I stop and learn how to write. But what do best friends know? Quite a lot, as it turns out. Edward Fenton kindly confirmed that my wild first draft was far from publication-ready.

Enter Debi Alper and Emma Darwin and the cast of the wonderful Self-Edit Your Novel course: Jo, Isabelwriter, Cathybramly, Peter, Woolybeans, Kath, Britta, Jane, Planetstalker and Jaxbees. I learned so much from all of you.

Cue tearing up of first novel and starting again.

Writing can be a lonely business; I couldn't have continued without the support of the writing community on the Word Cloud. Too many of you to name, but a special thanks to Moira and Jane for 'The Place to Write', to Alan Peabody for the annual short story competitions that introduced me to a whole new discipline, to William Angelo for maintaining a great tradition, to Ian Barnard for the game of tag and to the marvellous Randoms (brilliantly curated by the talented and generous writers Matt Willis and J. A. Ironside) for keeping me sane.

Hilary Dalenian, an inspiring leadership coach, and the WIG group – Pauline, Sarah, Hannah, Mitzi and Theresa – rebuilt my confidence and introduced me to the J. K. Rowling talk on failure. An exultation of professional editors knocked me back on track at various points: amy@thewritehelper, Jackie Buxton, Kate Foster and Rebecca Carpenter. Andrew Wille gave me encouragement and a book list and Debi Alper was always there when I needed her.

To my early Beta Readers, Christine, Shell, Barny, Britta and others expunged from memory through embarrassment – my sincere apologies. I had much to learn.

To my later Beta Readers, including Karen Ginnane, Jane Shufflebotham, Martin Gilbert, Ian Andrews, Gail Jack, Rupert Baines – thanks for your insight and generosity; with special mentions to Claire Barnard for her attention to detail, R. J. Jefferson for the mercenaries, Dr Raine Wilson for the flora and fauna of *polesia.*

Expert chemical help came from Ken Patterson and Tony Fishwick, foreign language assistance from Carmo, Barry and Erin, yachting advice from Jo Morgan and outdoor fun with River Deep Mountain High.

All mistakes are my own.

To Christine and Charles and the Greenough family – thanks for the Lakeland writing retreat.

To my wonderful in-laws, Pam and Andrew, who welcomed me into the Erskine family – thanks for the love and books.

To my children, Andrew and Joseph – thanks for the musical education and the excuse to go kayaking, sailing, skiing and surfing.

To Juliet Mushens, my fabulous agent, for believing in this project, and to Jenny Parrott, my awesome editor, to Jenny Page, Harriet Wade, Margot Weale, Thanhmai Bui-Van, Paul Nash and all the team at Oneworld, thanks for making it happen.

And to Jonathan, for his wisdom, patience, food and love. SYN.